ANALYSIS
INSTRUMENTATION

an introduction

ANALYSIS INSTRUMENTATION

an introduction

By

R.P. KHARE, Ph.D
Assistant Professor, instrumentation Centre
Birla Institute of Technology & Science,
Pilani (Rajasthan), India.

CBSPD

CBS Publishers & Distributors Pvt Ltd

New Delhi • Bengaluru • Chennai • Kochi • Kolkata • Lucknow • Mumbai
Hyderabad • Jharkhand • Nagpur • Patna • Pune • Uttarakhand

**Analysis Instrumentation
An Introduction**

ISBN: 978-81-239-0247-0

First Edition : 1993
Reprint: 2001, 2002, 2006, 2007, 2008, 2010, 2011, 2012, 2015, 2016, 2018, 2019, 2023

Published by Satish Kumar Jain and produced by Varun Jain for
CBS Publishers & Distributors Pvt Ltd
4819/XI Prahlad Street, 24 Ansari Road, Daryaganj, New Delhi 110 002, India
Ph: 011-23289259, 23266861, 23266867   Website: www.cbspd.com
Fax: 011-23243014   e-mail: delhi@cbspd.com
Corporate Office: 204 FIE, Industrial Area, Patparganj, Delhi 110 092, India

Ph: 011-4934 4934   Fax: 011-4934 4935   e-mail: publishing@cbspd.com; publicity@cbspd.com

Branches

- **Bengaluru:** Seema House 2975, 17th Cross, KR Road, Banasankari 2nd Stage, Bengaluru 560 070, Karnataka, India
  Ph: +91-80-26771678/79   Fax: +91-80-26771680   e-mail: bangalore@cbspd.com
- **Chennai:** 7, Subbaraya Street, Shenoy Nagar, Chennai 600 030, Tamil Nadu, India
  Ph: +91-44-26680620, 26681266   Fax: +91-44-42032115   e-mail: chennai@cbspd.com
- **Kochi:** 42/1325, 1326, Power House Road, Opp KSEB, Power House, Ernakulam, Kochi 682 018, India
  Ph: +91-484-4059061–65   Fax: +91-484-4059065   e-mail: kochi@cbspd.com
- **Kolkata:** 147, Hind Ceramics Compound, 1st Floor, Nilgunj Road, Belghoria, Kolkata 700 056, West Bengal, India
  Ph: +91-33-25633055–56   e-mail: kolkata@cbspd.com
- **Lucknow:** Basement, Khushnuma Complex, 7-Meerabai Marg (behind Jawahar Bhawan), Lucknow 226 001, India

  Ph: +91-522-4000032   e-mail: tiwari.lucknow@cbspd.com
- **Mumbai:** PWD Shed. Gala no. 25/26, Ramchandra Bhatt Marg, Next to JJ Hospital Gate no. 2, Opp. Union Bank of India, Noorbaug Mumbai 400 009, Maharashtra, India

  Ph: +91-22-66661880/89   e-mail: mumbai@cbspd.com

Representatives

- **Hyderabad** 0-9885175004 • **Jharkhand** 0-9811541605 • **Nagpur**   0-9421945513
- **Patna**   0-9334159340 • **Pune**   0-9923910676 • **Uttarakhand** 0-9716462459

Printed at SRK Graphics, Delhi, India

21st Century
Advent of Bright Future

Dedicated
To
Param Poojya Sat-Gurudev
Vedmurty, Taponishtha, Yugdrashta
Pandit Sriram Sharma, Acharya

PREFACE

The development of new materials requires a very precise investigation of their chemical composition and their physical properties. Such a material characterization often requires the use of sophisticated techniques and can be carried out only in advanced laboratories. The manufacture of materials may not require any qualitative investigation but a close control of product quality is always essential. The instrumentation requirement of the production unit, therefore, is that of routine or quantitative on-line analysis and process control. Measuring systems are also required for environmental monitoring and safety management. Hundreds of techniques have been developed for instrumental analysis and many more are being developed. The available books that deal with these techniques and related instrumentation, are too bulky. In fact, they have been written as reference books, meant primarily for the specialists in this field.

This book aims at discussing the fundamental aspects of analysis instrumentation, keeping in view the requirement of a student or that of a beginner in this field. Instead of discussing a large number of techniques of analysis and related instrumentation, the idea is to train the reader in arriving at the design of an analysis instrument (given the essential principle of measurement) by taking examples of some common analytical techniques. With this aim the book has been divided into ten chapters. Starting with the generalized configuration (Chapter 1), various off-line analysis instruments, that are commonly employed in analytical laboratories or in process analysis are discussed in chapters 2 through 9. The instruments dealt with are arc/spark emission spectrometers (chapter 2), UV/VIS/IR spectrophotometers (Chapter 3), flame emission spectrometers and atomic absorption spectrophotometers (chapter 4), fluorescence spectrometers (Chapter 5), X-ray analysers (Chapter 6), NMR spectrometers (Chapter 7), mass spectrometers (Chpater 8) and analytical electron microscope (Chapter 9). On-line analysers for industrial analysis and control, alongwith sampling systems are treated in Chapter 10. Analytical applications of these instruments have been purposely described, in brief, as various specialist works are available on this topic. The reader may refer to the bibliography cited at the end of this book.

In order to simplify the text, a modular approach has been adopted for describing the configuration of instruments. Each chapter (from 2 to 9) begins with an essential principle of measurement, that is followed by the criteria for design. The components are then described, in brief, to be followed by the discussion of design patterns. At the end of each chapter, a number of design-based problems, with sufficient hints, are given in the form of a tutorial. The electronic modules for signal processing have not been discussed at all for two reasons. Firstly, the students of B.E. (instrumentation) or even the students of science opting for this course would have normally gone through the courses like circuit theory, analog electronics, digital electronics, microprocessors, etc, and hence the repeatition of their contents is not justified. Secondly, for a beginner, the treatment of the subject like electronics would have doubled the present size of the book and so also its cost. Those who are interested in the details of electronic modules are respectfully referred to those many excellent books that are already available on this subject, though while studying this book, the need of any such reference book will not be felt.

For the past several years the author has been engaged in teaching and course development of analysis instrumentation, and other courses of similar nature; e.g. process analysis instrumentation, instrumentation technology, instrumental methods of analysis, etc. at Birla Institute of Technology & Science, Pilani. He was also involved in conducting almost a similar course in the fifth BITS- CSIR labile semester for the Scientists-B from different laboratories of CSIR. This book has grown out of this experience. The contents are more than enough for a one semester course on analysis instrumentation. The book will prove to be an invaluable asset to the B.E. students having a course on analytical instrumentation in prefinal or final year; or M.Sc. students of chemistry, physics or biology, having a course on instrumental analysis. Scientists and engineers involved in materials investigation and/or industrial analysis and control will also find it very useful.

The author wishes to thank all those who directly or indirectly contributed to making this venture a success. Sincere and specific mention must be made of Dr. S. Venkateswaran, Director, BITS, Prof. I.J.Nagrath and Prof. K.E. Raman for the encouragement; Prof. R.K.Patnaik, Prof. G.P.Srivastava, Prof. L.K.Maheshwari and Dr. G. Raghurama for providing all the facilities, Mr. Jamna Dhar Saini for wordprocessing the manuscript and Mr. K.N. Sharma for preparing the diagrams in the shortest possible time. Thanks are also due to my students who were very helpful and cooperataive during the preparation of the text. I am also indebted to M/S CBS Publishers and

Distributors, for timely publishing this work . Finally, I must thank my wife, Dr. (Mrs) Manjula Khare and my children, Gunjan and Amit, for their patience and cooperation during the entire course of this work.

R.P.KHARE

CONTENTS

1
A GENERALIZED CONFIGURATION

1.1 INTRODUCTION

An analysis instrument is a device or a set of devices that acquires the desired information regarding the chemical composition or the physical properties of a given sample or the process. This information may be required for a variety of purposes; e.g., testing of materials, maintenance of standards, verification of physical phenomena, monitoring the process stream, controlling the product quality, safety management and so on. Analysis Instrumentation is the science and technology of developing such measuring devices.

What criteria should be formed in deciding the configuration of an analysis instrument for a particular measurement? The answer is simple. If we want to know about the sample we need to find some way of communicating with it; that is to say, we need to find some way of interacting with the sample. The best way of interaction is to generate an appropriate signal and impose it onto the sample. The signal, after interaction with the sample, is attenuated or modified in accordance with some physical principle. This attenuated signal carries the measurement information. In order to present this information in a suitable form, some kind of signal processing may also be required. Accordingly, the functional elements in a generalized configuration of an analysis instrument may be arranged as shown in Fig.1.1.

There are many ways in which the same sample can be interacted and the measurement information obtained. These different ways consititute what are known as "analytical methods". Some common methods alongwith the way of interaction and the form of measurement information obtained are listed in Table 1.1. The selection of an analytical method for a particular measurement depends on the property that is suitable for analysis.

A brief discussion of the functional elements, that follows, will help us in understanding the generalized configuration depicted in Fig.1.1.

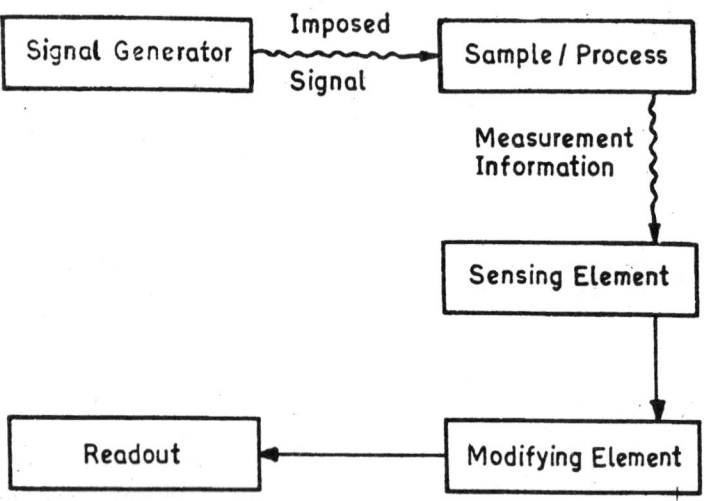

Fig. 1.1: Functional elements of a generalized analysis instrument.

Table 1.1: Some common techniques of analysing a given sample

Analytical method	Method of interaction with the sample	Form of measurement information
(a) Emission spectrometry	Excitation energy given to the sample by arc/ spark generator.	Optical radiation emitted by the sample
(b) Absorption spectro-photometry	Optical radiation(in uv,visible or ir range) is passed through the sample.	Attenuated radiation
(c) Flame Techniques		
-FES	Sample excited by the flame	Optical radiation emitted by sample
-AAS.	Radiation from hollow cathode source passed through the sample in flame.	Attenuated radiation
(d) Fluorimetry	Sample excited by external radiation	Optical radiation emitted by sample

(Contd.)

Analytical method	Method of interaction with the sample	Form of measurement information
(e) NMR spectrometry	Radiation generated by a r f- source	R.F. radiation absorbed by sample
(f) X-ray spectrometry	Primary x-rays excite the sample	Secondary x-rays emitted by sample
(g) Electron microscopy	Electron beam strikes the specimen	Various signals are generated

1.2. FUNCTIONAL ELEMENTS

1.2.1. Signal generator

It has been pointed out that an analytical measurement is a record of a response of the analytical species (i.e., the sample or the process) to an imposed signal; and hence a signal generator is a basic element of most of the analysis instruments. Conventionally, a module generating optical signals is called a "radiation source" or simply a "source" and that generating the electrical signals is called a "generator". The spectrum of the signal generated by the source may contain several frequencies. Therefore, to judge the response of the sample or the process over different frequencies, some kind of frequency isolation device is almost always required. Depending on the design requirement, the latter may precede the sample or immediately follow the sample .

1.2.2. Sensing element

In general, the signal whether it is generated by the sample (as in the case of emission analysis) or is altered after its interaction with a sample (as in the case of absorption analysis) is to be detected and transformed into a suitable form (often into an electrical signal) by a subassembly known as a detector or an input transducer. For example, a photodiode can detect the optical signals and also transform them into electrical signals.

In order that the sensing device be useful for measurement, it must respond to a given input parameter in a prescribed manner. In other words, the sensor output must be a measurable mathematical function of the input information. System's calibration generally establishes this functional relationship between input and output information. It is also important that the sensor does not respond to other inputs. If this is not the case, the extraneous inputs must be eliminated. The accuracy and dynamic response should also be considered while selecting a sensor for any instrument.

1.2.3 Modifying element(s)

A modifying element is generally required to perform three major functions; namely

(i) to change the input signal (which is the output of the sensor) to a suitable form;

(ii) to isolate the extraneous signals from the input signal; and

(iii) to amplify the input signal, if it is weak.

In some cases, further processing or modification, e.g., differentiation, integration etc. may also be required. However, every analysis instrument may not necessarily contain a modifying element. Its function may be performed by either the sensor or the readout device. For example, if a photomultiplier tube is used as a sensor, it performs three functions; namely, it detects an optical signal, transforms it into a photocurrent and amplifies it as much as 10^7 times.

It should be noted that the accuracy, range, and dynamic response of the modifying element should all be designed in such a way that suits both the preceding and the following modules; that is, the sensor and the readout device.

1.2.4 Readout device

The ultimate object of the analysis instrument is to present the information about the sample or the process in a form which can be interpreted by the human observer. Thus, the information carrying signal, after its modification to a suitable form is fed to the readout device, which may be a meter, recorder or other device. Simple examples of instrument outputs are pointer indications, digital readout and the pen positions on the chart of the recorder. The data can also be recorded photographically or stored magnetically.

1.3 DESIGN CONSIDERATIONS

1.3.1 Generalized design requirements

Whenever an analysis instrument based on a particular analytical method is to be developed so as to achieve a specific accuracy in the measured result, the following reaquirements are to be satisfied.

(a) It must correctly embody the essential principle of measurement.

(b) The act of measurement must not affect the system being analysed.

(c) The building blocks of the instrument, which are often called modules or elements, must have appropriate input, transfer and output characteristics.

(d) The elements must be related to each other and to the objects (e.g. sample or process) in a way that suits these particular objects and the intended accuracy in the final result.

(e) The design must be simple.

(f) The cost of the instrument must be reasonable.

All these requirements can not be fulfilled simultaneously in any real design ; and hence a practical design is a compromise between ease and economy on one hand and the level of performance on the other. This view can be established if one investigates the basic design patterns of analysis instruments.

1.3.2 Design patterns

The design patterns of analytical instruments may be broadly divided into three categories. They are :

(i) single channel design, in which the energy fed to the system moves along a single path from one element to another;

(ii) a double channel design, in which the signal traversing the measuring channel is compared with that traversing a reference channel and

(iii) a multichannel design, in which the flow of energy is channelized in several directions depending on the requirement.

Fig.1.2. shows a single channel design of a generalized absorption spectrometer. Herein, the optical signal traces a single path from source to the sample and the measurement information moves along a single path from detector/transducer to the readout device. Depending on the precision required in the measurement (or more correctly, in the final result), the design of this instrument can be simple or complex. For example, if a precision of 1-3% is required, the instrument needs a continuous radiation source followed by a bandpass optical filter (for frequency isolation), a simple detector, an amplifier and the sensitive output meter. Such a system will be cheaper and its maintenance minimal. Whenever simple instruments of this type are adequate for measurements of interest and the samples are to be analysed occasionally it is to be preferred over more complex systems. However, if a precision of ± 0.1% or better is required, the design of a single channel device will have to be more complex. The modified design may incorporate a stable source, a monochromator in place of a filter, a sensitive detector of wide dynamic range, a stable high gain amplifier and/ or a precision readout device. (See tutorials 1.2 and 1.3 for modifications in the single channel design). If the instrument has this kind of builtin stability, its cost will obviously increase.

Further, if the measurements are to be performed with reference to a standard sample, the utility of a single channel design is limited, because its operation would require the calibration of the instrument before each measurement. Apart from being a time-consuming procedure, the variables that affect the measurement will not be under full control and may fluctuate in magnitude at a rate faster than that can be accomodated by this procedure. In such cases a double channel design containing a reference channel and a measuring channel appears desirable. It will provide a degree of accuracy and sensitivity that can not be obtained with a single channel design.

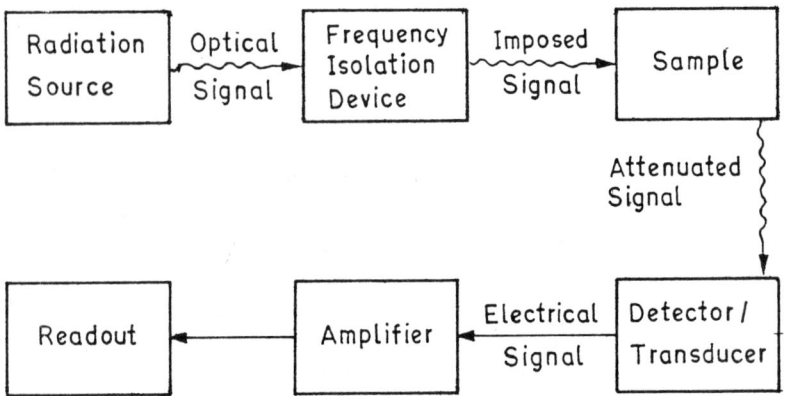

Fig., 1.2: A generalized absorption spectrometer

1.3.3 Operation and control

Apart from the basic layout, the mode of control of its operation is important in completing a particular design. There are three types of controls, namely :

(a) manual, in which the operator performs all the essential steps;

(b) automatic, in which the builtin closed control loops provide the controlling action and

(c) the computer control, in which the computer operates the instrument according to the instructions given to it.

In general, the control of a process variable means either to keep it constant or to vary it as a function of some other variable (known as standard). In both the cases, a signal indicating the deviation of the controlled variable from its setpoint, is necessary as an input to the controloer. A primary control element (also known as a sensing element) detects this deviation and sends the signals to the controller. The latter responds to it with an output signal to the final control element, which, in turn, produces the correction in the control agent. The process reacts to this correction, which is again detected by the sensing element and the input of the controller is modified accordingly. If there is no unbalance between the modified process variable and the standard variable, then the corrective cycle terminates.

However, if the unbalance persists, the cycle repeats itself. It is this feedback that modulates the correction and hence a closed control loop is sometimes called a feedback control loop. Different stages of such a loop are illustrated in Fig. 1.3.

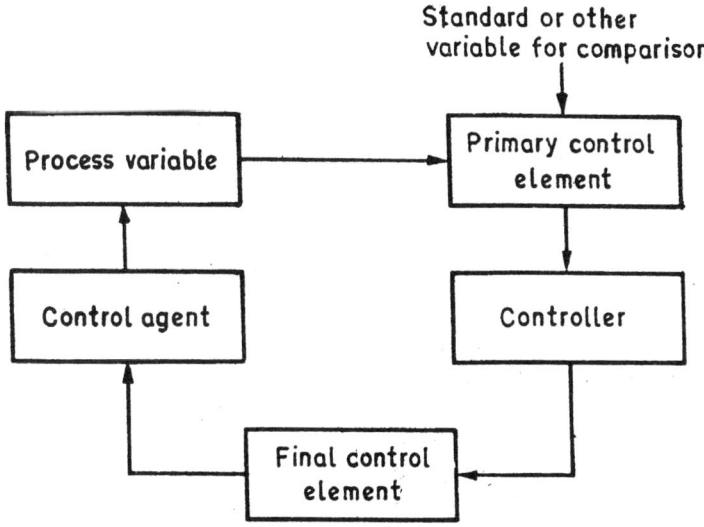

Fig.1.3: Closed control loop

Appreciable redesigning becomes necessary when a manually operated instrument is made to control its operation automatically.

1.3.4 On-line analysis

In modern industrial plants, it has become a tendency to install instruments which give immediate and continuous recording and control of product quality. An on-line analysis involves the measurement of the properties of the product at the point of production as opposed to removing the samples from the process from time to time and analysing them in a laboratory.

Broad features of an on-line analysis instrument are depicted in Fig.1.4. It consists of

(a) sampling system, which obtains the sample from the main process stream and carries it to the analysis instrument (if no prior preparation of the sample is needed) or to another instrument system where the sample is further processed to the required physical and chemical

state without removing its essential ingradients,
(b) analysis instrument, where the processed sample is analysed and
(c) the disposal system, where, after the analysis has been performed, the sample is disposed of. In some cases, the sample is returned to the main process stream at a suitable point.

The development of an on-line analysis instrument may require a high degree of skill on the part of the designer as well as that of an user, as this type of instrumentation may have a high degree of complexity. As the level of sophistication in the design increases, the speed, precision and versatility in measurement is gained but at the same time the initial investment as well as the maintenance cost of the instrument increases rapidly.

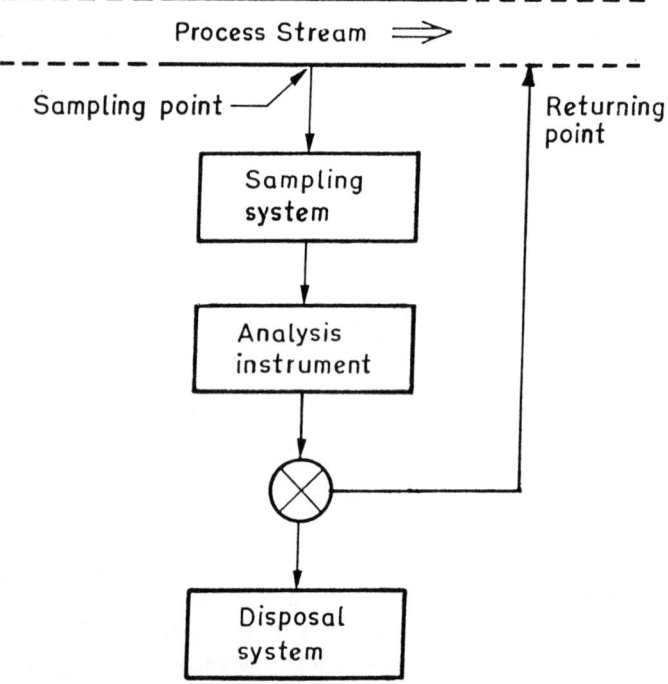

Fig. 1.4: Broad features of an on-line analysis instrument

1.3.5 Signal to noise ratio

The detector/transducer, in an analysis instrument, responds to the input signal and produces, generally, an electrical signal that carries the measurement information encoded as voltage, current etc. This transformed signal is amplified, processed as required in the context of application and is used to drive the readout. A careful observation of the electrical signals invariably reveals the presence of ripples or the spurious signal components that are unrelated to the phenomenon being measured. The ripples which may occur at any frequency within the capabilities of the system are called "noise". It is one of the goals of the instrument design to minimise noise or to optimise signal to noise (S/N) ratio, so as to achieve the desired sensitivity in the measurements.

Some knowledge of the origin of the noise may be helpful in understanding the measures that can be adopted to optimise the S/N ratio. The environment near the instrument may cause some disturbance. For example, the presence of electrostatic or electromagnetic field, mechanical vibrations, major fluctuations in the power systems and so on, are known to contribute to the noise. This type of noise is called an "environmental noise" and may be eliminated, at least in principle, by proper design and other measures taken during the measurements. Other types of noise; e.g., thermal and shot noise, arise because of the discrete nature of matter and energy and are continuous and spontaneous. These are, therefore, not subject to elimination, even in principle. The thermal noise is the voltage, having its origin in the thermally induced random motion of charge carriers, whereas the shot noise is the voltage or current arising from the random times of release of charges. For example, the random arrival of photons at a photodetector gives rise to a random fluctuation in the photocurrent. In fact, the shot noise is an inherent property of photodeteactors and even of the optical signal itself.

Accordingly, the extraction of information from noisy signals becomes a problem whenever low signal intensities are to be handled. What measures exist for improving the sensitivity of measurement in such cases ? Besides enhancing the signals and reducing the noise by all possible means, there exist various approaches for improving the S/N ratio. One of the methods amounts to extending the period of observation during which the signal is effectively averaged and hence the random noise component is minimized. In another approach, the extraneous signals are filtered from the analytical signal. In some cases, where the frequency of noise is different from that of the analytical signal, modulation and demodulation of the signal helps in reducing the noise.

1.4 PERFORMANCE CHARACTERISTICS

The performance characteristics of an analysis instrument are indicative of its capabilities and limitations for a particular application . They can be broadly divided into static and dynamic characteristics. When a process variable changes rapidly, the dynamic relations between the input and output of the instrument are governed by relevant differential equations. In a number of cases, though, the input to the instrument may be constant or vary quite slowly. In such cases, it is not necessary to be concerned with dynamic descriptions. Thus various static performance parameters like, repeatability , accuracy, uncertainty, resolution etc, are usually good enough to give a meaningful description of the instrument. It should be remembered though, the overall performance of the instrument is judged by both the static and dynamic characteristics. In the following lines, we discuss in brief, a few common terms that are employed in evaluating the performance of analysis instruments.

1.4.1 Repeatability

The term refers to the degree of resembalance in a set of measurements of the same quantity measured under the same conditions by the same observer employing the same instrumental setup. For example, if the indicating meter of a given instrument shows ± 2% fluctuation for a group of measurements of the same input (no fluctuation in the input signal) under the same conditions, the repeatability of the instrument is ± 2%.

1.4.2 Accuracy

This term refers to the closeness of the instrument output to the true value of the measurand (as per standards). However, it is, generally specified as an uncertainty or inaccuracy in measurement from the true value. The inaccuracy may be caused by a number of factors, such as temperature, vibration, drift, hysteresis and so on. The drift, which is the variation of the instrument output, is generally caused by the component instability. Similarly, the hysteresis is caused by backlash error, friction or the presence of magnetic materials.

1.4.3 Resolution

It may be defined as the smallest increment of the measurand that can be detected with certainty by the instrument. In spectroscopic systems, the distinction should be made between the terms resolving power and resolution. While the resolving power is entirely determined by the optical design of the monochromator, the observed resolution depends on the resolving power, the source intensity, monochromator transmittance, detector sensitivity and the degree of perfection of other optical and electronic components.

1.4.4 Sensitivity

The static sensitivity, which is sometimes termed as incremental gain or scale factor, is defined as the ratio of the change in magnitude of the instrument output to the corresponding change in the magnitude of the measured quantity (i.e. the input).

1.4.5 Linearity

It is desirable that the sensitivity of the instrument remains constant for all values of the measurand. In other words, the indicating scale should be linear. The linearity, however, is measured in terms of the maximum deviation from linearity (generally as percentage of full scale).

TUTORIAL-1

1.1 For an unknown sample, there can be two types of analysis; viz, (i) qualitative analysis, which is an effort to know what is inside the sample (e.g., impurities, compositional ingradients, etc) and (ii) quantitative analysis, which involves measuring how much of that (.e.g, percentage of impurities etc) is there in the sample.

What change if any, should be made in the generalized configuration of Fig.1.1, if the instrument is to perform both types of analysis?

(It may be noted that for each constituent in the sample, we may need a separate signal).

1.2 The radiation source in the instrument shown in Fig.1.2, is run by a power supply. It is found that the stabilization of the input power to the source is inadequate for precision measurements. What modification in the design of this instrument is needed for uniform source intensities? Sketch and explain the modified design of the instrument.

Hint: A feedback loop connecting the source and the power supply may be formed. This will regulate the voltage output of the power supply for constant source intensity. Note that for a precision measurement other changes may also be neacessary.

1.3 Assume an optical slit is placed between the source and the sample in Fig.1.2. Devise a closed control loop for an automatic adjustment of the optical slit, so that light intensity remains constant. Explain the functions of different elements of the loop.

Hint: Note that the process variable here is the intensity of the light beam (falling on the sample) which is dependent on the width of the slit. If a part of the beam is diverted onto a second detector and the transformed signal is compared with a standard one, the deviations from the setpoint can be used to drive the slit in the direction of unbalance.

1.4 The development of new materials requires a thorough qualitative investigation of the samples. Such an investigation has to be carried out in a R & D laboratory. However, the control of the quality of the sample being manufactured may require measuring (quantitatively) only the deviation of some variable (e.g. concentration etc) from the setpoint.

What may be the similarities and differences in the design of off-line and on-line analysis instruments?

2

EMISSION SPECTROMETERS

2.1 INTRODUCTION

An elementary system such as an atom, an ion or a molecule, when reasonably isolated, appears to have a number of discrete energy states. Such a system may be forced into an unstable yet allowed energy state, i.e, the one above ground state, by means of thermal or electrical excitation or by exposure to a radiation of suitable wavelength. The excited species, e.g. an atom may loose excess energy either radiatively or non-radiatively as shown in Fig.2.1.

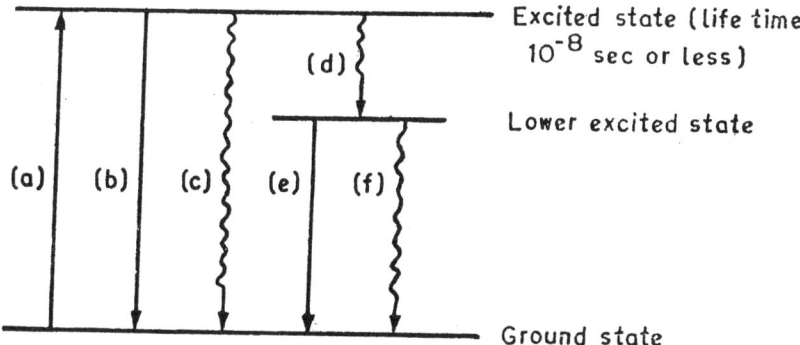

Fig.2.1: Generalized energy changes in an elementary system.
- (a) Absorption of energy
- (b) subsequent loss of energy: re-radiation
- (c) non-radiative dissipation of energy as heat
- (d) internal conversion to lower excited state and partial loss of energy as heat followed by either;
- (e) re-radiation or again
- (f) loss as heat

The radiative return transitions constitute the emission spectrum of the species. The line spectra are emitted by atoms and ions and the band spectra are emitted by molecules. In order to obtain a line or band spectrum of a particular sample (solid or liquid), it is converted into a vapour which is then energized by the excitation source. Since each element emits its own spectrum, it is possible to detect the presence of different elements in a given sample. This technique is reliable and can be used for qualitative as well as quantitative elemental analysis for concentrations as low as 1 ppm. The instrument which employs this principle of measurement is known as an emission spectrograph or an emission spectrometer.

2.2 DESIGN CRITERIA

What should be the criteria for designing an emission spectrometer? It becomes obvious from the above discussion that an emission spectrometer should provide for the vaporization (or volatilization) and excitation of the sample as well as for recording of the spectrum emitted by the sample. Accordingly, the modules in a generalized emission spectrometer may be arranged as shown in Fig.2.2.

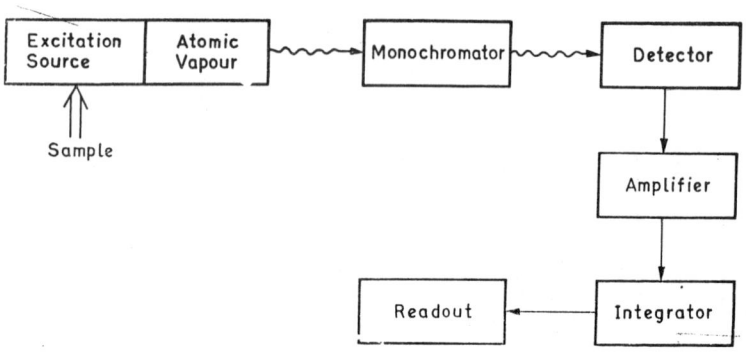

Fig. 2.2: Block diagram of a generalized emission spectrometer. Wavy arrows depict the optical signal and straight arrows represent the ele-ctrical signal.

The importance of each of these modules may be understood, if one investigates the dependence of the signal developed by the detector on pertinent variables.

It can be shown (*1,2) that the detector signal (S) may be expressed by a relation of the type given below.

$$S = \left[\frac{(1-\alpha)R_a}{v} \right] I_\lambda (c,T), M(\theta,\lambda). D(\lambda) \qquad (2.1)$$

Herein α is the fraction of analytical species that is ionized (and, therefore does not contribute to the signal); $(1-\alpha)$ represents the fraction that remains unionized; R_a is the rate of atomization or volatilization of the sample and v is the velocity of transport of the sample through the excitation zone. $I_\lambda (c,T)$ represents the intensity of a spectral line emitted by the atoms of the analytical species and is a function of the concentration, c, of the emitting atoms and the temperature, T. The intensity of an emission line is defined as the energy of wavelength, λ, radiated per second. It is equal to the product of the number of atoms of the analytical species spontaneously undergoing a given transition in a second and the energy of the photons (of wavelength, λ) released. $M(\theta, \lambda)$ is a geometrical factor that determines the solid angle(θ) of radiation observed by the monochromator and its transmittance with wavelength (λ) and finally, $D(\lambda)$ represents the response of the detector as a function of wavelength.

It is evident from eqn (2.1) that the bracketted quantity has a considerable influence on the detector signal. Consequently, the excitation source should be such that it maximizes R_a and minimizes α and v. Further, it should yield higher intensity, I_λ, of an emission line. The monochromator and detector should be such that their response is maximum as well as uniform in the range of emission spectrum. The analytical species may contain several elements that may not all be volatilized at the same time and hence the output of the detector should be integrated over the period of volatilization.While selecting the modules for an emission spectrometer, these criteria must be kept in mind.

2.3 EXCITATION SOURCES

The radiation source for an emission analysis is obtained by the excitation of the material under investigation. In principle, therefore, an excitation source should be such that it volatilizes and excites all the elements present in a given sample and should present a spectrum at an intensity that reflects the concentrations of the elements. Further, the excitation must be constant and

*(1) L.De Galan: "The posibility of a truly absolute method of spectrographic analysis": Analyt. Chem Acta 34,2 (1966).
(2) H.A. Strobel, in chapter 13 of "Chemical Instrumentation: a systematic approach to instrumental analysis." Second edition, Addison Wesley, Massachusetts (1973).

the source must transfer reproducible amount of energy to the sample. The latter is particularly important in quantitative analysis. These requirements can be met in a variety of ways and depend on the state of material to be analysed.

The volatilization of the sample can be brought about by the application of thermal energy and the excitation can be produced by either thermal or electrical energy. However, most excitation methods use both thermal and electrical energy. Such methods have been divided into three groups; namely

    (i) arc excitation,

    (ii) spark excitation and

    (iii) flame excitation.

Here we shall discuss first two types of methods only, the third type shall be taken up while discussing the flame photometers (in chapter -4).

2.3.1 Arc excitation:

An arc is developed between two conducting electrodes set at a distance corresponding to a break-down voltage*, when they are connected to a power supply. The latter allows a continuous electrical discharge of high intensity and extremely high temperature to pass between the electrodes. At such high temperatures, the gas molecules and atoms are split into ions and electrons to form an arc-plasma. Some of the typical characteristics of an arc are as follows.

    (i) The discharge column is contracted in front of the electrodes, so that the so-called 'burning spot' in front of the electrodes has a much smaller area than the average cross-section of the arc plasma. The spot wanders over the electrode area.

    (ii) The plasma energy is concentrated near the cathode. As a consequence, the temperature near the cathode becomes very high (of the order of 6,000 to 10,000° C). For this reason, the material to be analysed is placed on the cathode or is introduced near the cathode where it is quickly vaporized and gets excited.

    (iii) The voltage drop between the electrodes is not linear. Further, arc has a negative resistance characteristic, i.e., its resistance decreases as the current is increased.

In spectrochemical analysis, usually, vertically oriented free-burning arcs are employed. In the arc plasma, mainly the atoms are excited, so that the

---

* For a particular arc gap (i.e., the distance between the two electrodes), there exists a certain minimum voltage, V_b, known as break-down voltage, that is necessary to produce the arc between the electrodes.

spectrum of the emitted light consists mainly of atomic lines. However, a weak continuous background is also found to overlap the atomic lines.

2.3.2 Arc generators:

The equipment used for generating an arc is known as an arc generator. Depending upon the requirement of operation, it can be designed in several ways. The basic circuit for a d.c. continuous arc generator is shown in Fig.2.3. It contains a bridge rectifier whose output is smoothed by a capacitor of appropriate capacitance (C), an ammeter A, a variable resistor R, for adjusting the arc current and a voltmeter (V) for checking the terminal voltage of the power supply. In high voltage arc generators, the arc is self-ignited (if the terminal voltage is greater than the break-down voltage of the arc gap), whereas in the mains voltage arc generators, the ignition of the arc is achieved either mechanically or by the use of Tesla current. The latter is a high voltage, high frequency and low consumption current. The mechanical ignition consists of bringing the two electrodes momentarily into contact and then seperating them. Normally, mains voltage d.c. continuous arc generators are employed for excitation purposes. The d.c. mains voltage ranges from 110 to 220 volts and operating current varies from 3 to 15 amps. This technique is quite sensitive and the analytical species (i.e. elements) that are present in low concentration can easily be detected. The reason for this is that the emission spectra consists of relatively few, easily excited atomic lines. However, during arcing, the electrode tips are strongly heated which leads to 'fractional distillation' of the front part of the electrodes. As a result, reproducible intensity of spectral lines can not be achieved.

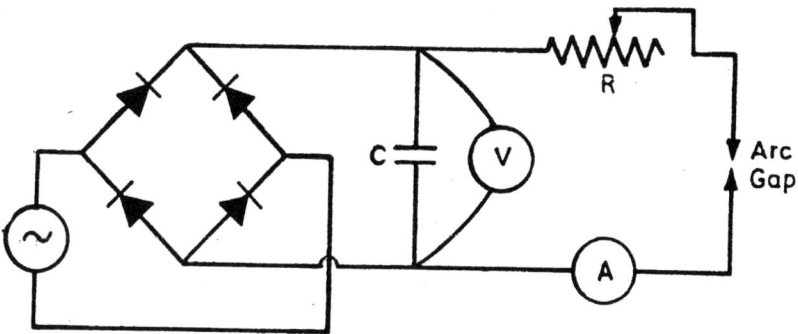

Fig.2.3: Circuit for d.c. continuous arc generator

This difficulty is partially removed with mains-voltage a.c. arc generation. While applying the a.c. mains supply, it should be kept in mind that the voltage curve reaches zero twice and the polarity of the electrodes is inverted during one cycle. Thus an a.c.arc generator can become self supporting only when the operating voltage is higher than the breakdown voltage of the arc gap; otherwise it will need ignition current every half cycle. Accordingly, the circuit for mains voltage a.c. arc generators consists of two parts (a), an arc current circuit and (b) the ignition circuit (or a Tesla current generator). The two circuits are shown in Fig.2.4 (a and b).

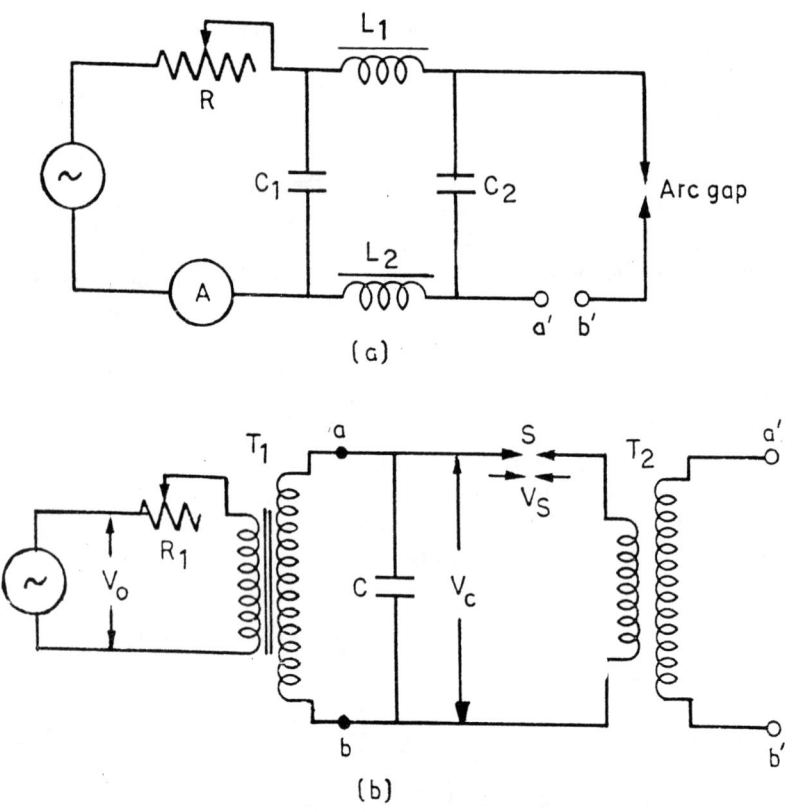

Fig. 2.4: Mains voltage a.c. arc generator:
    (a) arc current circuit,
    (b) ignition circuit.

In a Tesla generator (shown in Fig.2.4 b) the mains voltage (V_o) is stepped up to about 5 - 10 KV through a buffer resistance R_1 and the transformer T_1. This voltage charges the capacitor C (of about 1000 pF). The charged capacitor discharges through the primary coil of the Tesla transformer T_2 connected in series with the secondary spark gap (S). The capacitor discharges in that phase instant when the voltage V_c becomes equal to the break down voltage, V_b, of the secondary spark gap (i.e., $V_c = V_s = V_b$). The discharge of the capacitor is a high frequency damped oscillatory phenomenon. The frequency of damped oscillations is determined by capacitance C and the inductance, L, of the primary coil of the Tesla transformer. (Generally, the d.c. resistance of the circuit is negligible and hence the frequency, in Hz, is given approximately by $\dfrac{1}{2\pi}\sqrt{\dfrac{1}{LC}}$). The corresponding frequency is of the order of 10^7 Hz . This high frequency current in the primary of T_2 induces a high frequency current in its secondary. The output of the latter is used for ignition. Following the discharge, the capacitor C is recharged and this process of charging and discharging continues until the voltage V_c drops below the level of breakdown voltage of spark gap, S(as shown in Fig. 2.5). The duration of charging and the number of charges and discharges in a half cycle are determined mainly by the values of R_1 and C and on the electrode distance in the spark gap, S.

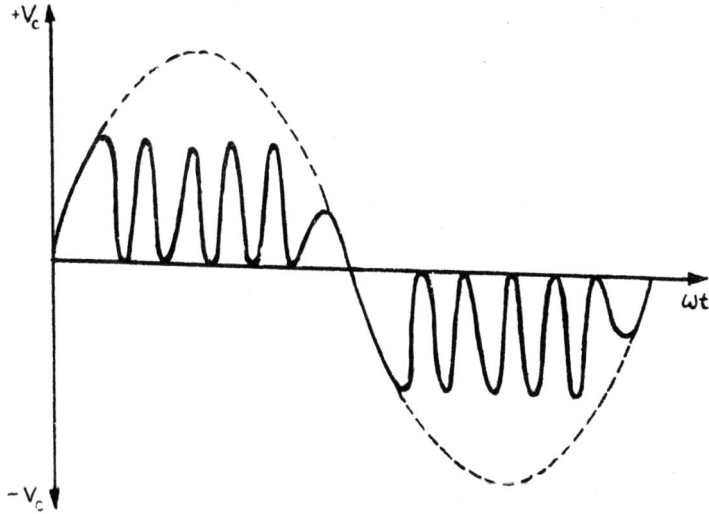

Fig.2.5 : V_c versus t curve (schematic), for a Tesla current generator. Dashed curve shows the form(not magnitude) of variation of V_o with t.

Once the arc is ignited the Tesla current is taken over by the working arc current. In the circuit of Fig. 2.4 (a) the arc current also passes through the secondary of the Tesla transformer connected to the points a' and b', and the ignition current passes through the arc gap via the capacitor C_2. The latter has practically no impedance towards high-frequency current, although it blocks 50 Hz mains current. The impedance of the capacitor is lowered when the frequency ω is increased according to the relationship $R_c = 1/j\omega\ C$. In the arc current circuit, a L-C filter consisting of two inductors L_1 and L_2 and a capacitor C_1 has been employed in order to prevent the high frequency ignition current from getting into the mains. The impedance of an inductor increases with frequency according to the relation $R_L = j\omega\ L$. The inductors L_1 and L_2, therefore, allow free passage of the 50 Hz arc current, while allowing only a small fraction of the high frequency current to pass. The latter is short circuited through the capacitor C_1 and is thus not admitted into the mains.

A drawback in the above system is that the arc is ignited in that phase instant when the voltage of capacitor C (in the Tesla generator) reaches the breakdown level of the spark gap S. The value of this breakdown voltage is generally uncertain. It depends upon, among other things, the condition of the front part of the electrodes of the spark gap. Therefore, it is not possible to control the "phase interval of the ignition current very precisely". This disadvantage is eliminated if the ignition current is either mechanically controlled by replacing the spark gap S in Fig. 2.4b by a rotary gap; or, electronically controlled through a circuit shown in Fig. 2.6.

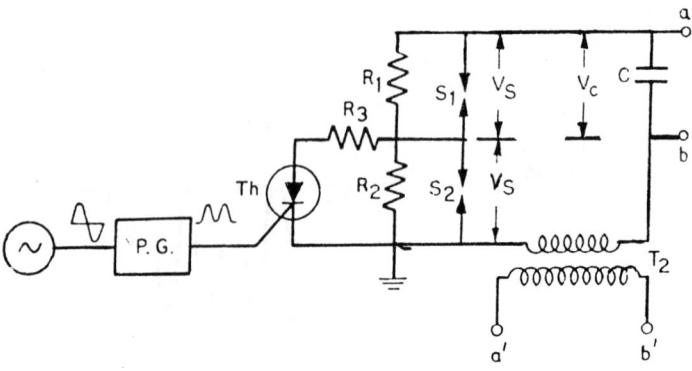

Fig. 2.6: A popular Bardocz circuit for electronic control of ignition current. Herein the points a, b and a'b' refer to Fig. 2.4b.

In this case, the controlling spark gap S of Fig. 2.4(b) is replaced by two identical spark gaps S_1 and S_2 connected in series; and both spanned by two resistors R_1 and R_2. The control signal is obtained through a pulse generator (P.G.) which sends a positive pulse in each half-cycle. This pulse ignites the thyristor (Th) and makes it conducting if its anode is at a sufficient positive potential. The potential drop, V_c, across the capacitor C should be such that

$V_c > V_s$; where V_s is the breakdown voltage of spark gap S_1 or S_2; and, at the same time, $V_c < 2 V_s$. The charged capacitor C discharges at that instant when the thyristor receives a positive pulse. The latter renders the thyristor conducting and the anode current starts flowing through the resistor R_2. Therefore, the spark gap S_2 is short circuited and a large potential drop appears across S_1. The latter thus breaks down. As a consequence, larger potential difference now appears across the resistor R_2 and therefore S_2 also breaks down. In this way, the capacitor discharges through both the spark gaps S_1 and S_2. This discharge is again a high frequency damped oscillatory phenomenon and therefore, gives rise to a high frequency current at the terminals of the tesla transformer T_2. After the discharge, the potential difference across R_2 reduces so much that the anode current is terminated and the condition that prevailed before the discharge is restored.

2.3.3 Spark excitation

A spark is a quickly terminated electrical discharge between the two electrodes, when the potential difference between them is greater than the breakdown voltage of the electrode gap. The electrical discharge leading to spark is produced by the charged particles that are always present in the space between the electrodes. On the application of the electric field, the charged particles get accelerated and ionize the atoms with which they collide. This process first occurs immediately in front of the electrodes and later spreads over the whole electrode gap. In a very short time, the number of charged particles becomes very large. The space between the electrodes thus becomes conducting and the spark proper is produced.

The electrical discharge that can properly be called a spark is produced when a charged capacitor is discharged through the electrode gap.

2.3.4 Spark generator

The electrical devices, that are used for generating the sparks, are known as spark generators. They may be classified according to the charging voltage of the capacitor or according to the method of control. The circuit generally employed for a high voltage (10 - 30 KV) spark generator is given in Fig.2.7. The capacitor C is charged by a mains voltage a.c. through a damping resistance R_d and transformer T, and it is discharged through the inductance L, Ohmic resistance, R and the analytical gap (AG). The capacitor discharges at that instant when its terminal voltage becomes equal to the breakdown voltage of the analytical gap. This discharge is a damped oscillatory phenomenon and lasts for about 10^{-7} to 10^{-3} sec. Following the discharge, the capacitor is recharged. In a single half-period, the charging and discharging occurs several times.

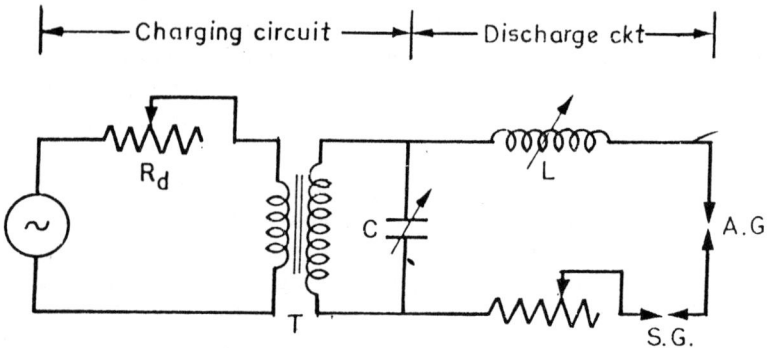

Fig. 2.7:Circuit diagram for a high voltage spark generator

In the non-controlled spark generators, the breakdown voltage varies during sparking thereby altering the maximum current of the capacitor discharge. As a consequence, the temperature of the spark plasma is changed, which causes a corresponding change in the intensity of spectral lines. The reproducibility of spark generation can be improved considerably if a second spark gap (S.G.) is also included in the discharge circuit. The second spark gap, known as the controlling gap is made up of tungsten or platinum electrodes. The distance between these electrodes may be reproducibly set employing a micro-meter screw. Further, they are water-cooled and a constant air stream blown into the electrode gap removes the excess charged particles left after the discharge of thecapacitor. In this way, the breakdown voltage of the controlling spark gap(S.G.) is maintained constant. In another version of control, known as Feussner system, the controlling gap is replaced by a motor driven rotating switch.

There are other varieties of the spark generators, where the terminal voltage is quite small; for example, about 1 KV for medium-voltage and 110-230 V for low-voltage or mains - voltage spark generators. In these generators,the terminal voltage is insufficient to produce a breakdown of the spark gap, and hence, the capacitor discharge is initiated by a high frequency, high voltage ignition current.

2.3.5 Excitation by d.c. plasma jet

Plasma jet is a flow of partially ionized gas that is blown out of a small orifice. This gas stream has a flame like structure. A d.c. plasma jet is essentially a stabilized d.c. arc. It consists of two graphite electrodes placed one above the other and surrounded by a water-cooled chamber as shown in Fig.2.8. A d.c. current of about 20 amp is maintained in order to establish an arc between the two electrodes. The arc plasma is blown out through the hole in the upper electrode by inert carrier gas to produce a plasma flame. The sample is introduced into this flame by the sprayer, also called a nebulizer or atomizer. The function of secondary electrode is to reduce the movement of the so-called cathode spot on the cathode surface.

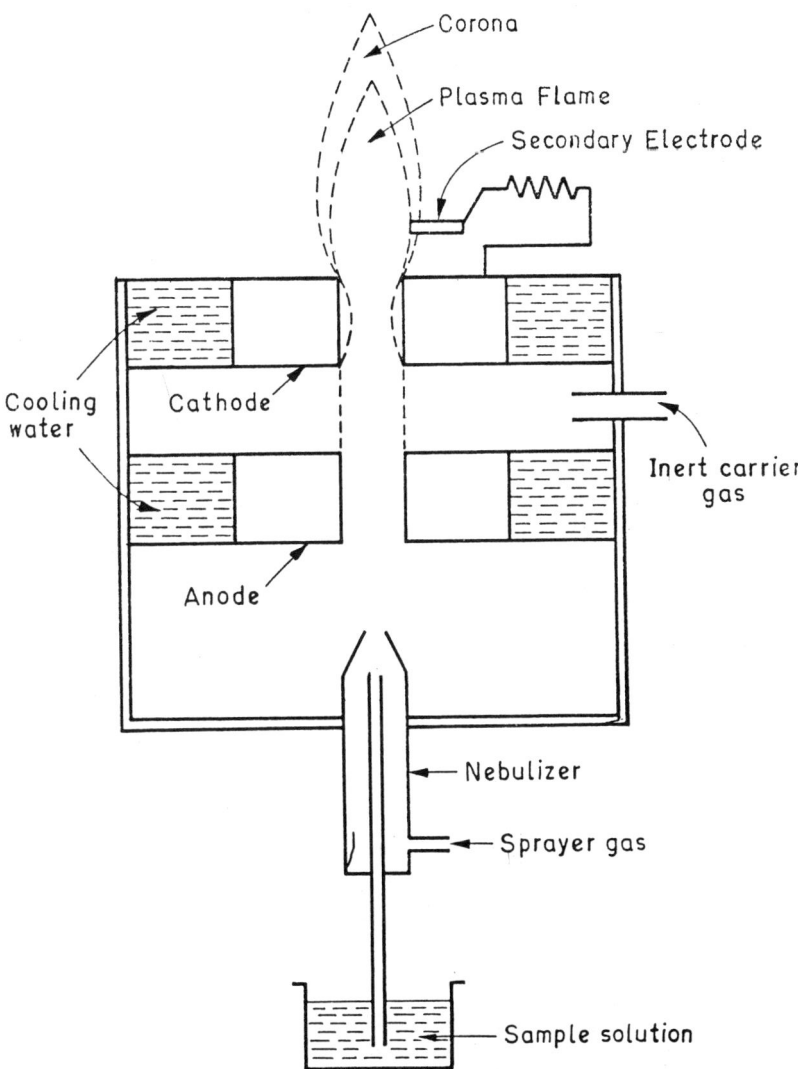

Fig.2.8: Schematic diagram of a plasma burner

The circuit diagram of the power supply for the plasma jet is shown in Fig.2.9. Here the current is rectified employing a single diode and the pulsations are smoothed by the capacitor C. The arc is initiated by a high frequency, high voltage ignition current. The choke L is effective only in igniting the arc.

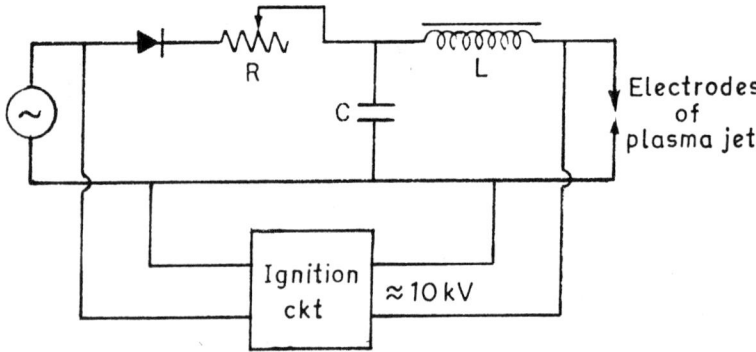

Fig.2.9 Circuit diagram of the power supply for a d.c. plasma jet

It is instructive that the carrier gas should have high excitation potential so that its spectrum is not observed. Generally, an inert gas e.g. argon or helium is used for this purpose. With proper gas flow rate and the size of the hole in the upper electrode, very good analytical reproducibility may be achieved.

2.4 MONOCHROMATORS

A monochromator is an important module of most of the spectroscopic analysis instruments. It is a device for separating the constituent spectral components of a beam of polychromatic radiation. It's essential components are :

   (i) an entrance slit, for providing a narrow optical image

   (ii) a collimating device (e.g. lens or mirror), which makes the rays, spreading from the slit, parallel;

   (iii) a dispersion device (e.g. a prism or grating) which produces an angular separation between different wave lengths present in the incident beam of radiation;

   (iv) a focussing device (again a lens or mirror), which reforms the images (proportional to the size of the entrance slit) of the separated wavelengths at the exit curve of the monochromator; and

   (v) an exit slit, for isolating the desired spectral wavelengths. A generalized configuration of a monchromator is shown in Fig.2.10.

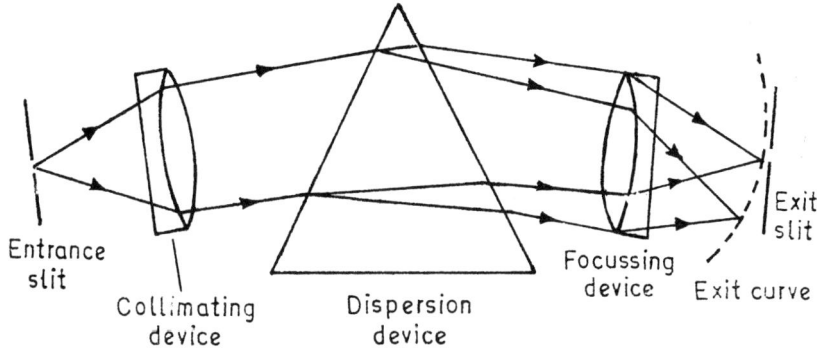

Fig. 2.10: A generalized monochromator showing essential components

2.4.1 The slit system

Physically, an entrance slit is a narrow rectangular aperture, which admits light into the monochromator; but, virtually, it behaves as a source of radiation for the optical system of a monochromator. If it is illuminated by a source of monochromatic radiation, the optical system forms a single image of the entrance slit at the exit slit. If it is illuminated by a polychromatic source (as shown in Fig.2.10), the image of the entrance slit corresponding to all the spectral consituents (i.e. bands or lines) present in the polychromatic radiation will be formed in the focal plane, also known as exit curve, of the focussing device and these images can be brought (usually by rotation of the dispersion element) onto the exit slit (which is kept stationary at some point in the exit curve). Thus the ability of a monochromator to distinguish one spectral band (or line) from another depends not only on the dispersion element, but also on the size and intensity of the image and hence on the slit width, the illumination of the slit and the design of the slit. We shall take up the design aspect first.

Slit design: The slit is generally formed by two slit jaws which may be fixed or whose separation may be changed. In the case of adjustable slits, either the position of one jaw can be changed, the other remaining fixed (unsymmetrical slits) or both the jaws can be moved with repect to the slit centre (symmetrical slits). It is essential, in both the fixed as well as adjustable slits, that the edges of the slit jaws be straight and parallel in order to get a clear image. Further, the edges must be sharp in order to avoid the reflection of light from the jaws into the monochromator. For adjustable slits, the jaws must remain strictly parallel in different settings of the slitwidth, (i.e.when the slitwidth is changed). It is also desirable that the jaws return to identical setting on readjustment. Only a careful design and good mechanical work can fulfill these requirements.

In general, the slit jaws are made of a metal that is non-corrosive and capable of taking a sharp edge, and high polish. Stainless steel, stellite and nickel are often used. Ocasionally metal coated quartz plates are also used. For adjustable slits, the guidance of the slit jaws constitute a basic design problem. Whatever be the mechanism in general, it has been found that the slit is guided by adjusting a screw or a cam and closed by a spring of some sort. This spring is interposed between the guiding agent and the carriers of the slit jaws. The guiding screws have micrometer heads with which the slit width can be determined. Generally, the slit is covered by a quartz or glass plate in order to protect it from dust etc. In some cases, slit lens or field lens serves a similar purpose, in addition to focussing the light on to the slit or the collimator lens. In spectrographs, mounts are provided in front of the slit to carry Hartmann diaphragm for limiting the length of the slit.

The slit width: The choice of mechanical slit width (which is the separation of the two slit jaws) depends largely on the work to be performed by the monochromator. When the slit width is larger, the spectral lines or bands become more intense but wider and hence the resolution is decreased. The opposite is true when the slit is narrowed. In general, the slit width is so large that diffraction effects due to the slit are negligible; however, if a greater resolution is required, narrower slits maybe employed. Faust has prescribed the following guidelines for the setting of optimum slit width which is defined by $S = 1.3 \ k\lambda$ (where k is the f-number of the collimator lens and λ is the wavelength of light):

(i) the intensity of the spectral lines at this width should be about three fourths of its maximum possible intensity,

(ii) the width of the spectral line should be approximately $1.25 \ k'\lambda$ where k' is the f-number of the camera and

(iii) the resolution limit should be approximately 80% of that attainable at an extremely narrow slit.

Slit Illumination: The flux of light that passes effectively through a monochromator (also known as the energy throughput of the device) becomes important particularly when weak spectral lines are to be analysed. If the energy throughput is more, the signal to noise ratio of the instrument will be enhanced and hence, a greater precision in the measurement will be obtained. Given a light source, how to optimise the energy throughput of the monochromator? The energy throughput of a monochromator is largely dependent on the light gathering power of the collimator lens provided the slit is not too narrow. Therefore, if the slit is illuminated in such a way that the light gathering power of the collimator lens is completely utilized the energy throughput can be optimised. This can be achieved by imaging the light source either on the slit or on the collimator lens. The source is imaged on the slit, if the structure of a definite part of the source is to be examined and it is imaged on the collimator lens if its structure is to be omitted from the spectroscopic lines.

2.4.2 Collimating and focussing devices

The function of a collimating device is to render parallel the rays of all the wavelengths present in the radiation passed through the slit. With actual lenses, it is not possible to get exact parallelism of the beam of radiation for all wavelengths and apertures, the accuracy depends largely on the lens design, and the lens materials. The design of a collimator lens and a camera or a focusing lens offers almost the same problem because the function of one is exactly the reverse of the other. The difficulty arises because of two major aberrations caused by the lenses. The chromatic aberration, arises from the fact that the focal length of a lens is a function of refractive index and hence of the wavelength; consequently, light rays of different wavelengths can not be focused at the same point. Even for a single wavelength the rays passing through different zones of the lens can not be focussed exactly at a single point. This fault is known as spherical aberration.

There exist certain measures, for correcting these faults. For example, achromats and apochromats are combined lenses composed of two or three materials and are corrected for chromatic aberration of two and three colours respectively; aspherical lenses are generally free from spherical aberration and so on. Even with best possible design, it is not possible to remove all the aberrations and correct for all the wavelengths. Consequently, most of the sophisticated instruments employ mirrors rather than lenses for collimating and focussing purposes. The mirrors are generally of the front surface type i.e. a sheet of glass or silica is ground and polished to the desired curvature and a highly reflecting metal or metal-dielectric film is then deposited on its face.

The action of a lens as well as mirror can be described approximately by the thin lens equation :

$$\frac{1}{P} + \frac{1}{Q} = \frac{1}{F} \qquad (2.2)$$

where P and Q are object (here a light source) to lens (or mirror) and lens (or mirror) to image distances respectively and F is the focal length of the lens (or mirror).

2.4.3 Dispersion devices

The design and the mode of operation of a dispersion device is of fundamental importance in the working of a monochromator. The dispersion of a polychromatic radiation involves angularly separating the different wavelengths present in the wavefront. It can be accompalished either through the mechanism of refraction by the prism or through diffracation by the grating.

Prism systems

The prism systems are employed to serve a variety of purposes in the design of optical systems. For example, wedge scanning prism, isosceles right triangular prism, penta prism, leman Spenger prism, Amici-prism, porro prism, rhomboidal prism etc. are principally designed for changing the image orientation or the direction of the optical axis by reflection; Abbe's reversion prism, dove prism, and pechan prism systems are used for the rotation of image through its inversion ; and; Nicol, Glan Foucault; Rochan and Wollastan prism are used for producing polarized light and so on.

The utility of a prism as a dispersion device arises from the phenomenon of refraction. When a ray of light falls obliquely on the interface between two media of refractive indices n_1 and n_2 $(n_2 > n_1)$ it is refracted i.e., gets deviated from its original direction at the interface (see Fig.2.11 a) in accordance with the snell's law:

$$\frac{Sin\ i}{Sin\ r} = \frac{n_2}{n_1} \tag{2.3}$$

where i and r are the angles of incidence and refraction respectively. If the medium 2, is a plane parallel plate, then the emergent ray will exhibit a parallel displacement but no net angular deviation is produced. If the faces of the plate are not parallel but are arranged as in a prism (as shown in Fig.2.11 b), an useful deviation of a monochromatic ray can occur. Since the refractive index of the prism material varies with wavelength, an useful isolation of different wavelengths present in the incident radiation can occur; thus causing the prism to act as a dispersion element.If the refractive index of the material of the prism decreases with increasing wavelength, then the prism shows a normal dispersion which is the mode depicted in Fig.(2.11.b). This mode is commonly used in analysis instrumentation.

Incident Ray

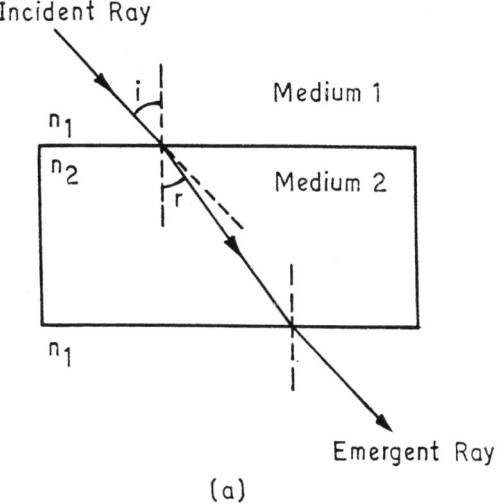

(a)

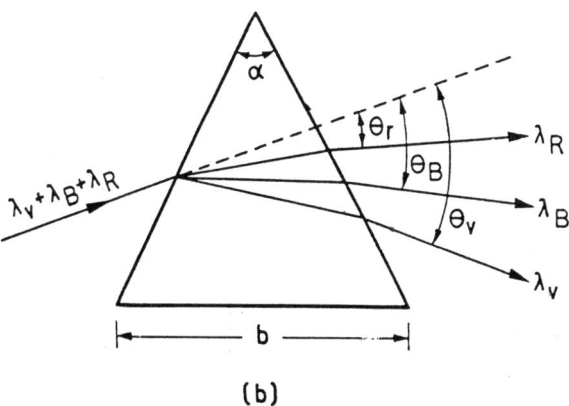

(b)

Fig.2.11: Schematic representation of

(a) Refraction by a plane parallel plate
(b) Dispersion by a prism: α is the apical angle of prism and b is the base length

The angular dispersion of a prism is defined as the variation in the angle of deviation (θ) with wavelength (λ), at constant angle of incidence. In other words, it is a measure of the angular separation ($d\theta$) of two rays that have a wavelength difference of $d\lambda$. It is expressed as $d\theta/d\lambda$.

It can also be written as

$$\frac{d\theta}{d\lambda} = \frac{d\theta}{dn} \times \frac{dn}{d\lambda} \tag{2.4}$$

where the term $d\theta/dn$ depends on the geometry of the prism and is called a geometric factor and the term $dn/d\lambda$ is characteristic of the prism mataerial and hence it is called a specific factor or a characteristic dispersion. For a minimum deviation position (i.e. a position in which the ray passing through the symmetrical prism travels parallel to its base and hence the angle of deviation of the emergent ray is minimum), it can be shown that

$$\frac{d\theta}{dn} = \frac{2 \sin (\alpha/2)}{[\ 1-n^2 \sin^2(\alpha/2)]^{1/2}} \tag{2.5}$$

where α is an apical angle of prism. There is no such corresponding formula for $dn/d\lambda$ except that computed through

Hartmann's equation:

$$\frac{dn}{d\lambda} = - \frac{c}{(\lambda-\lambda_o)^2} \tag{2.6}$$

where c and λ_o are so called Hartmann's constants. This formula is valid for small ranges of wavelength.

Further, assuming the minimum deviation, the resolving power R of the prism is given by :

$$R = \frac{\wedge}{d\lambda} = \frac{b\ dn}{d\lambda} \tag{2.7}$$

where b is the base length of the prism. It should be noted that this expression for R, assumes that the entire prism face is illuminated. If this is not the case, then b should be replaced by (t-s) where t and s are the thickness of the prism through which the two extreme rays of the beam pass.

Prism designs and mounts

As the prisms are generally used at or near the position of minimum deviation, with equal incident and emergent angles, the common prism systems are based on the isosceles prism. What should be the apical angle ? It directly follows from equation (2.5) and some consideration of the reflection losses. However, 60° prism angle has been found to be the optimum for most materials and hence equilateral. prisms are most common.

Prisms of anisotropic materials; such as quartz, are cut in such a way that the crystallographic axis runs parallel to the light rays. In order to avoid circular birefringence, a 60° prism is made of the two symmetrical 30° prisms, one of dextrorotatory and other of lavorotatory quartz, as shown in Fig.2.12. In this way, the double refraction produced by the first half of the prism is cancelled by equal and opposite effect on the other half. Such a prism is known as Cornu prism.

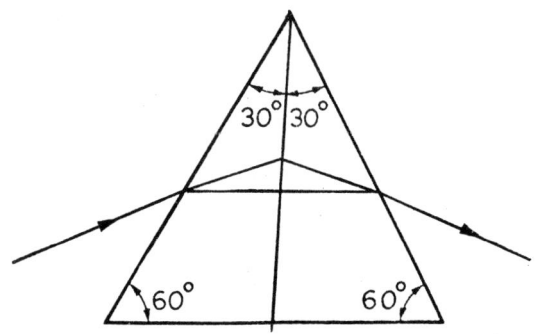

Fig.2.12: Cornu Prism

A 30° prism or half prism is employed in Littrow mounting shown in Fig.2.13. This design of the prism monochromator is widely used because (i) it exhibits a greater dispersion in a compact arrangement; and (ii) a single lens or mirror serves as both the collimating as well as focussing device. Here, as the light passes through the prism, it gets refracted and dispersed. The dispersed rays suffer reflection at the back surface and travel back through the prism, so that a 30° prism effectively work as 60° prism.

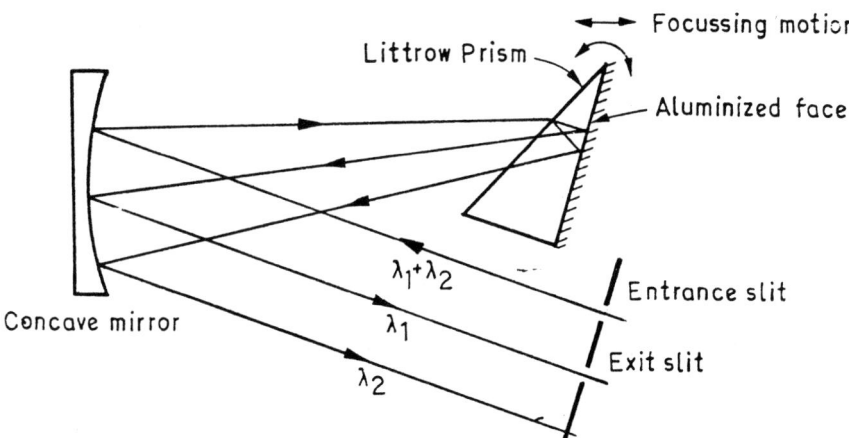

Fig.2.13 : Littrow mount, Schematic. (There are also other possible configurations employing a littrow prism)

A constant deviation prism e.g. a Pellin-Broca prism (shown in Fig.2.14), is a four-sided prism in which the incident beam suffers two refractions at surface AB and AD and a total internal reflection at surface BC. Such a system permits the dispersion with a constant deviation for any wavelength (transmitted by the material of the prism). With the help of a Pellin Broca type of prism (which is equivalent to two 30° half prisms combined with a right angled totally reflecting prism.), a fixed deviation of 90° between the incident and the emergent beams can be achieved. When a poly-chromatic beam enters through AB at an angle of minimum deviation, only one of its wavelength strikes the surface BC at an angle of 45°. That wavelength, upon emergence from the prism gets deviated by just 90° with respect to the incident beam. Therefore, by rotating the prism (preferably about the point 0 in Fig.2.14), different wavelengths can be deviated through 90° and hence can be brought on to the exit slit (not shown).

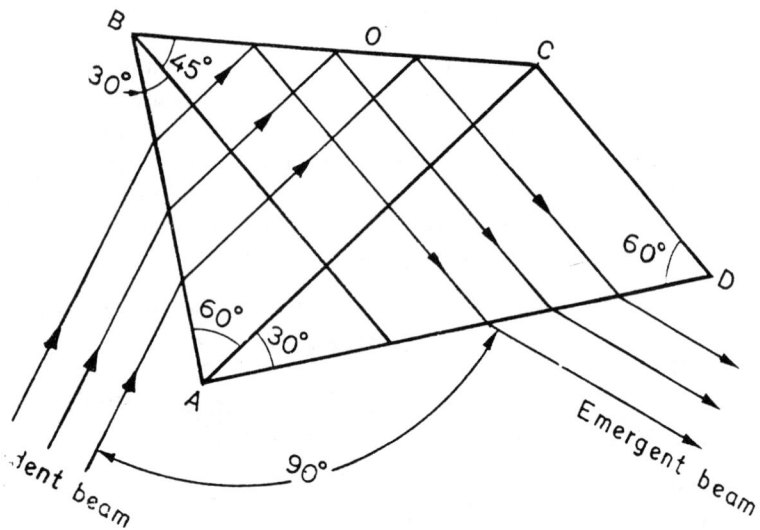

Fig.2.14 : The deviation of monochromatic light in a Pellin- Broca Prism

Another arrangement giving constant deviation is shown in Fig.2.15. This is known as Wadsworth mount. A light beam, after being deviated (and dispersed in the case of polychromatic beam), by the prism, is reflected by the mirror which is placed on the extended line of the base of the prism. At the minimum deviation, the incident and the emergent beams are parallel. The wavelengths can be scanned by rotating the mount about the point 0.

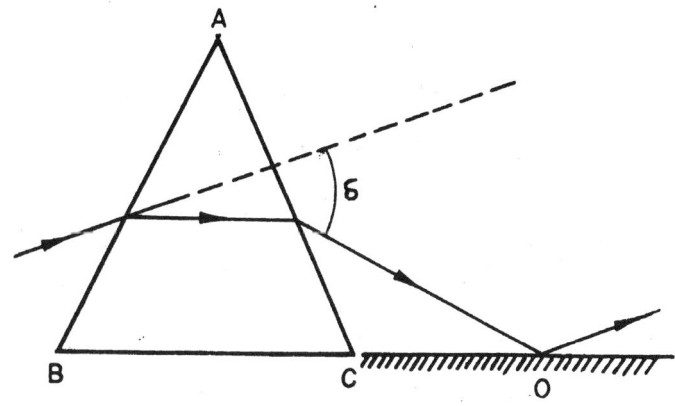

Fig.2.15 : Wadsworth mount (schematic)

Grating Systems

It is important to note that, for a prism monochromator, the dispersion decreases with increase in wavelength. This effect , generally, limits the resolving power of such monochromators, at the red end of the spectrum. Moreover, in the ultraviolet below $0.12\,\mu m$ and in the infrared beyond $40\,\mu m$, suitable transparent materials are not available for making prisms. These problems are easily overcome, if the prism is replaced by a grating.

What is a grating and how it works? When a parallel beam of monochromatic radiation falls normally on a slit of narrow width W, and the radiation transmitted by the slit is focussed on a screen, the image of a slit is diffraction pattern (which consists of a central maximum with a series of minima and maxima on either side) as shown in Fig.2.16. The width of the central maximum is approximately $2F\lambda/W$, where F is the focal length of the lens and λ is the wavelength of the radiation. The width is then inversely proportional to W. If W is so small as to be of the order of λ, the central maximum may become so wide as to cover the entire field of view. A diffraction grating is made up of such narrow slits arranged to form an array of parallel, equi-distant and closely spaced slits. In a plane transmission grating, the light passes through these slits; in a reflection grating, the light is reflected from the surface that has been ruled with a large number of parallel grooves so that each groove behaves as a slit. In both the cases, the fundamental phenomenon is the same and hence an example of transmission grating may be sufficient for understanding the action of gratings.

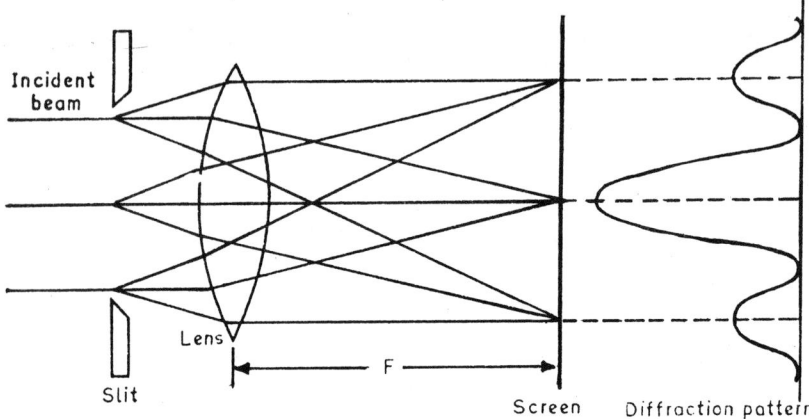

Fig.2.16 : Fraunhofer diffraction by a single slit

Plane transmission grating : If the individual slit width is so narrow that the central maximum (obtained in the diffraction pattern of a single slit) fills the entire field of view, then new maxima and minima are found (in this wide central maximum) arising from the interference between beams of radiation passing through different slits. With a large number of slits, these new maxima become sharp and narrow as compared to the distance between them. It is these maxima that constitute the spectra of a grating.

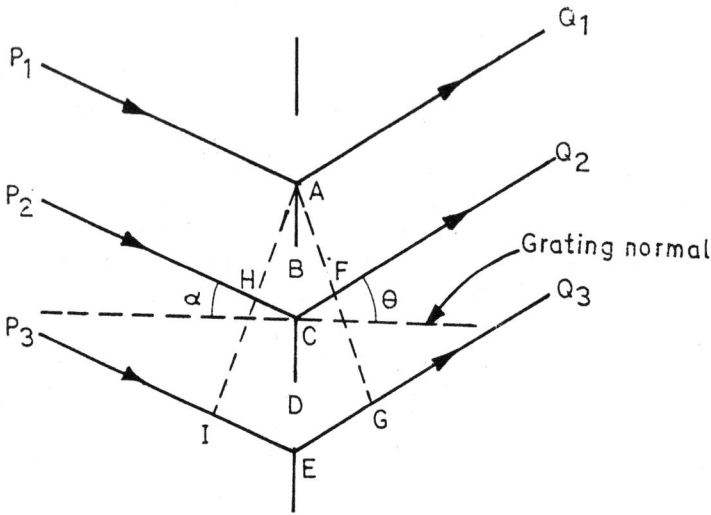

Fig.2.17 : Diffraction by a plane transmission grating

The position of the maxima (due to a monochromatic beam incident on the grating) can be determined employing the geometrical arrangement shown in Fig.2.17. In this figure, ABCDE represents a section of the grating where BC and DE are the adjacent openings and AB and CD are the opaque spacings between the openings. A beam of parallel rays P_1 P_2 P_3 is incident on the grating (from the left) at an angle α with the grating normal. After passing through the grating, the diffracted light moves out from each opening in all directions, as the openings are sufficiently narrow. The condition for constructive interference, between rays from different openings can be obtained by considering a set of parallel rays labelled Q_1 Q_2 Q_3, diffracted at an angle θ with the normal.

In the incident beam, the rays P_1, P_2 and P_3 are in phase at AHI; where AHI is perpendicular to the direction of propagation of the rays. Likewise, AFG is perpendicular to the direction of propagation of the diffracted rays (Q_1, Q_2 and Q_3). It is obvious that the ray $P_2 Q_2$ travels a path greater than the ray $P_1 Q_1$ by a distance HCF, to reach AFG plane, and in the same fashion, the ray $P_3 Q_3$ travels a distance IEG greater than $P_1 Q_1$. It is obvious from Fig.2.17 that IEG=2 HCF and that the path difference for succeeding openings will increase in arithmatical progression.

If the rays Q_1, Q_2 and Q_3 are brought to a focus, and if the path difference, HCF is an integral multiple of the wavelength, of the incident radiation, these rays will reinforce one another and produce a bright image of the source. Thus if AC = CE = d (i.e., if the grating element = d), then bright maxima will occur for conditions which satisfy the relation

$$HC + CF = d\ (\sin \alpha + \sin \theta) = m\lambda$$

where m is an integer (1,2,3 etc), called the order number and θ is the angle of diffraction (measured with respect to the grating normal). Since the condition for reinforcement can also be met by rays diffracted on the opposite side of the grating normal the complete expression for the image position is

$$d(\sin \alpha \pm \sin \theta) = m\lambda \qquad (2.8)$$

By convention, positive sign will be used in the left hand side of eqn 2.8, if α and θ, both, are on the same side of grating normal and negative sign will be used when they are on opposite sides of the normal. For normal incidence, i.e., when $\alpha = 0$, the eqn. (2.8) reduces to a simple form, $d \sin \theta = \pm m\lambda$.

Eqn. (2.8) is a fundamental grating equation and it holds for all values of α and θ and for both transmission as well as reflection gratings. It should be noted that in a grating spectrum, the shortest wavelength is deviated least, where as the opposite is true for a prism spectrum. The angular dispersion of a grating may be obtained by differentiating eq. (2.8), assuming α to be constant;

$$\frac{d\theta}{d\lambda} = \frac{m}{d \cos \theta} \qquad (2.9)$$

It is obvious from eqn. (2.9) that the dispersion is minimum, when $\theta = 0$, and the spectrum is observed on the grating normal. For this normal spectrum, $d\theta = (m/d) \, d\lambda$, or for small changes in wavelength; $\Delta\theta = $ (const.) $\Delta\lambda$ approximately. Therefore the dispersion is a constant and λ is a linear function of θ.

A troublesome feature of the grating is the overlapping of the spectra of different orders, when a polych-romatic light beam is incident on the grating. It is clear from eqn.. (2.8) that for fixed values of α and θ, if a wavelength λ is observed in the first order ($m = 1$), there will also appear, at the same angle θ, higher orders of other wavelengths such that $\lambda_1 = 2 \, \lambda_2 = 3 \, \lambda_3 = m\lambda_m$.

The following relations hold:

$$\lambda_2 = \frac{\lambda_1}{2} \, , \quad \lambda_3 = \frac{\lambda_1}{3} \, , \quad \lambda_m = \frac{\lambda_1}{m}$$

For example, the spectra of $0.60\,\mu m$ in the first, $0.30\,\mu m$ in the second and $0.20\,\mu m$ in the third order are observed at the sam, angle θ. These overlapping orders can be avoided in a straight-forward fashion. Generally, cut-off filters are employed to remove unwanted spectral ranges. Sometimes light is first dispersed by a prism and only a narrow spectral range of wavelengths is then allowed to fall on a grating. In latter arrangement, the movement of the prism is synchronised with that of the grating so that both respond to the same spectral range.

The linear dispersion $dl/d\lambda$, of the spectrum obtainable on the exit curve of the focussing device is given by the relation

$$\frac{dl}{d\lambda} = f \times \frac{m}{d \cos \theta} \qquad (2.10)$$

where f is the focal length of the focussing device.

The theoretical resolving power of the grating is given by the expression,

$$R = \frac{\lambda}{d\lambda} = mN \qquad (2.11)$$

where N is the total number of lines or grooves illuminated. Thus R is dependent on the product of the order and the number of grooves or rulings and not on the wavelength or the grating element, d.

Reflection grating

A reflection grating consists of an array of parallel grooves which are identical in depth and shape, equally spaced; and of a material that is totally reflecting and stable. Fig.2.18 depicts schematically a cross-section of such a grating. In spectrochemical analysis, reflection gratings of this type are often preferred, as by properly shaping the grooves, it is possible, to concentrate the available intensity at a particular diffraction angle. A grating of this type, in which the face is made up of plane facets lying at an angle, to the surface of the grating is known as an echellette grating and the angle ß is called as the angle of blaze. This grating is highly efficient in diffracting wavelengths close to those for which specular reflection occurs. In Fig.2.18, this case is represented for the incident and diffracted rays. If ϕ is the angle of incidence and reflection (with respect to the facet normal) then $\phi = \alpha - \beta = \beta + \theta$. Since θ may be positive or negative (i.e., it may be on either side of the grating normal), we can write

$$\beta = \frac{(\alpha \pm \theta)}{2} \qquad (2.12)$$

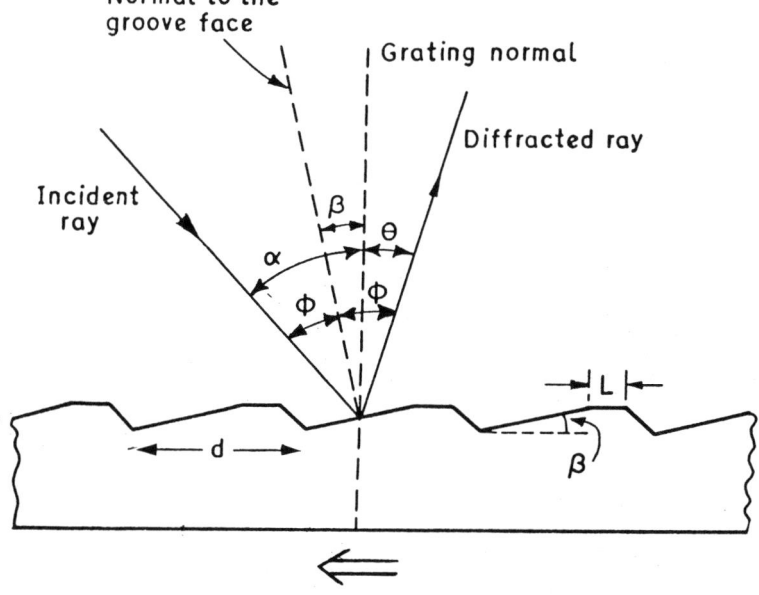

Fig. 2.18 : Cross-sectional diagram of a blazed reflection grating. ß is the blaze angle; α is the angle of incidence; θ is the angle of diffraction; d is the grating element; and L is land; the ratio of (d-L)/d is the aspect ratio. The arrow (⇒) shows the direction of blaze.

If the specular reflection and the m th order diffraction are coinciding for a particular wavelength, the grating demonstrates maximum efficiency at that wavelength. The latter is called a blazed wavelength, λ_B. It can be found from the grating equation (2.8), i.e. $m\lambda = d(\sin \alpha - \sin \theta)$ by putting $\lambda = \lambda_B$ and $\theta = \alpha - 2\beta$ (from eqn.2.12).

$$\text{Thus } m\,\lambda_B = d[\sin \alpha - \sin(\alpha - 2\beta)]$$

For normal incidence, $\alpha = 0$ and hence

$$m\,\lambda_B = d \sin 2\beta$$

$$\text{or} \quad \lambda_B = \frac{d \sin 2\beta}{m} \quad\quad\quad (2.13)$$

This equation generally appears in other forms, which can be shown equivalent (for example, see tutorial 2.1)

The performance of a reflection grating is evaluated by its efficiency, which is defined as the fraction of monochromatic light diffracted in a given order relative to the fraction specularly reflected by a flat mirror. It is obvious that a particular wavelength can be diffracted efficienctly in one order only. If the efficiency is higher, the grating becomes more useful, as it can ensure a greater energy throughput (to the monochromator) and less scattered light.

Concave grating

A concave grating combines the principle of plane grating and the focussing properties of a concave mirror. The grating law, (eqn. 2.8), describes its diffraction behaviour. An important characteristic of such a grating is that, it reduces the monochromator to three parts, an entrance slit, a concave grating, and an exit slit. Moreover, the chromatic aberrations and losses caused by the use of lenses are avoided, although, the errors of a concave mirror, particularly, astigmatism, can not be discarded. Accordingly, it is possible with a concave grating to investigate in the regions of far u.v. and i.r. which can not be tackled by plane transmission gratings.

Grating mounts

The grating systems have been classified according to the shape of the grating and the manner of mounting the gratings, entrance and exit slits, and mirrors relative to one another. Nearly all the mountings of the concave grating are adaptations of the Rowland's principle. It states that if a concave grating is placed tangent to a circle that has a diameter equal to the radius of curvature of the grating blank such that the grating centre lies on the circumference, the diffracted spectrum of a source (i.e., slit) placed on the circle will be focussed on the circumference of the circle, (see Fig.2.19). This circle is known as a "Rowland circle".

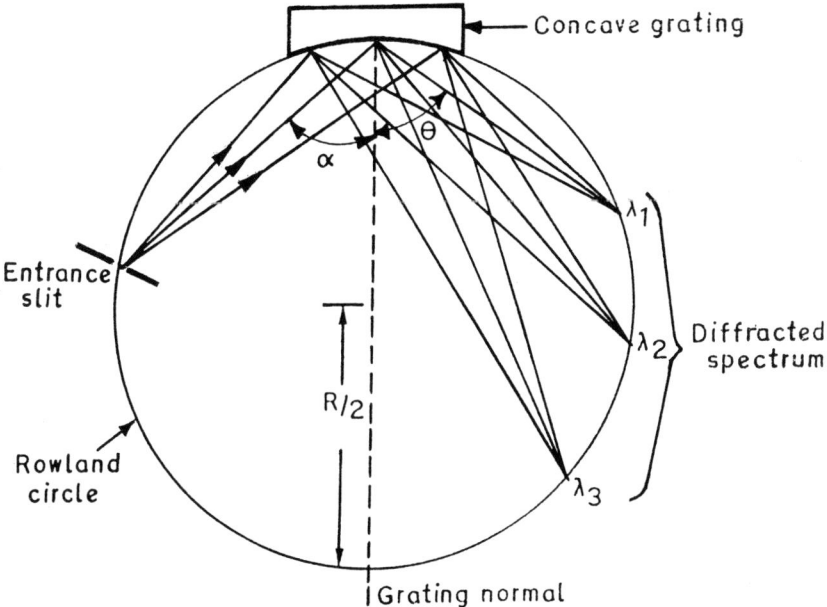

Fig. 2.19 : Rowland circle: A beam of radiation ($\lambda_1 + \lambda_2 + \lambda_3$) from the entrance slit is diffracted and focussed by the grating at λ_1, λ_2 and λ_3; α and θ are the angles of incidence and diffraction respectively, R is the radius of curvature of the grating.

Some of the important mounting systems for concave gratings are illustrated in Fig.2.20 to 2.23.

Appropriate mountings for concave gratings are bulky and require undue mechanical linkage complications. Moreover, large , accurately ruled and blazed plane gratings and replicas are now available at reasonable costs. For this reason, plane gratings are finding increasing applications in scanning spectrometers.

Littrow mounting, which has been discussed earlier (see Fig.2.13 for the prism mounts) can also be used with plane gratings. In fact, the plane grating can be directly substituted for the prism. This mounting has been successfully used in the infrared monochromators. Typical case studies may be seen in tutorial 2.1 and 2.4.

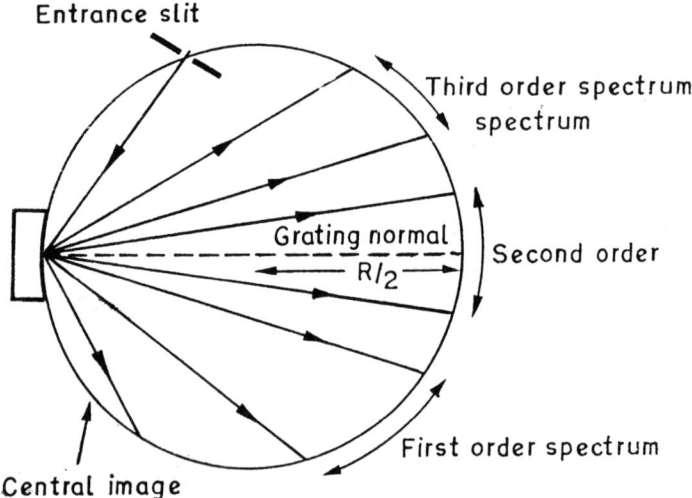

Fig.2.20 : Paschen Runge mounting: In this mounting the entrance slit and the grating are kept fixed on the Rowland circle and the exit slit(s) or the detector is mounted (or moved) on the circle for the spectral region of interest. This mount is generally used to screen vacuum ultraviolet region between 0.03 to 0.2 μm.

Fig.2.24 shows Fastie-Ebert mounting which is also compact. It utilizes a single concave mirror (M) for collimating as well as focussing. The spectrum is scanned by rotating the grating (G). Czerny-Turner mounting shown in Fig.2.25, is a modification of the Fastie-Ebert mounting. It utilizes two concave mirrors (M_1 and M_2) in place of a single large concave mirror. This arrangement has become more popular for scanning spectrometers where resolution and aperture ratio are prime requisites.

2.4.4 Scanning systems

The monochromators focus the spectrum of the source (i.e., the illuminated slit) at the exit curve. The spectral lines or bands present in this spectrum can be either recorded simultaneously by putting a photographic plate in the exit curve (and removing the exit slit) or scanned one by one by bringing successive lines or bands onto the exit slit. How the scanning of the spectrum can be implemented? In principle, it can be done by

    (i) moving the exit slit (i.e., the detector),

    (ii) moving the optical field and

(iii) a combination of the two systems.

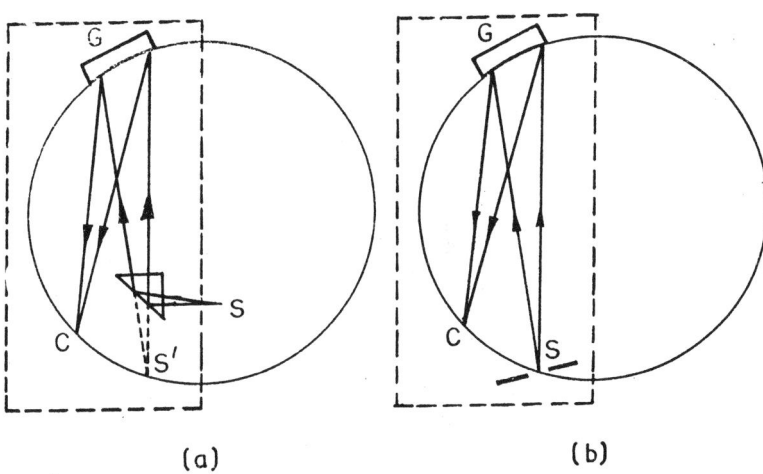

(a)   (b)

Fig.2.21 : Eagle mounting (a) off-plane mount in which the entrance
slit (S) is placed inside the Rowland circle and a total
reflecting prism(P) is employed to direct the beam of radiation
onto the grating from a virtual slit(s') on this circle. The
diffracted spectrum is focussed on the camera(c), (b) in-
plane mount, in which the slit and the camera are placed
side by side, on the Rowland circle. In order to change the
wavelength range, the grating (G) is moved along the optical
axis toward or away from the slit and is rotated about its
vertical axis. This mechanism of movement has the effect
of rotating the Rowland circle in space about the slit as the
axis. In Eagle mounting, since the angle of incidence and
angle of diffraction are nearly equal, it results in a compact
arrangement.

The first one is useful only for slow speed operation as it has a drawback
of subjecting the detector to mechanical vibrations that can increase the noise
of the transduced signals. The second is most practical and is implemented by
oscillating or rotating the prism or grating or the mirror.

For plane grating mounts, a simple sine drive mechanism (illustrated in
Fig.2.26). is generally employed.

It permits the wavelength to be read directly on the counter. This drive
provides a linear displacement (x) proportional to the sine of the angle, ϕ,
through which the grating (G) is rotated. When $\phi = 0$, the angles of incident
and emergent rays (α and θ respectively) with respect to the grating normal

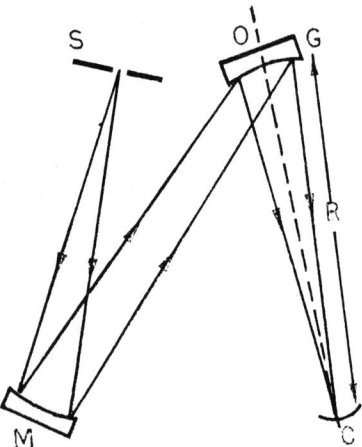

Fig.2.22 :   Wadsworth mounting The collimating mirror (M) illuminates
the grating (G), which can be rotated about the point 0. The
distance R of the camera (C) from the grating is increased
as the wavelength increases. To minimise aberration, the
slit(s) must be close to the grating.

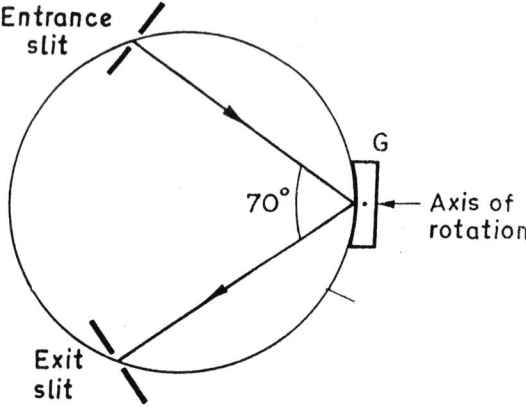

Fig.2.23 :   Seya-Namioka mounting: As the grating (G) is rotated about
a vertical axis through its centre the slits move from the
Rowland circle and the wavelength at the exit is increased.
In this arrangement an angle of approx. 70° is maintained
between the incident and the diffracted beams. This mounting
offers a simplest scanning mechanism.

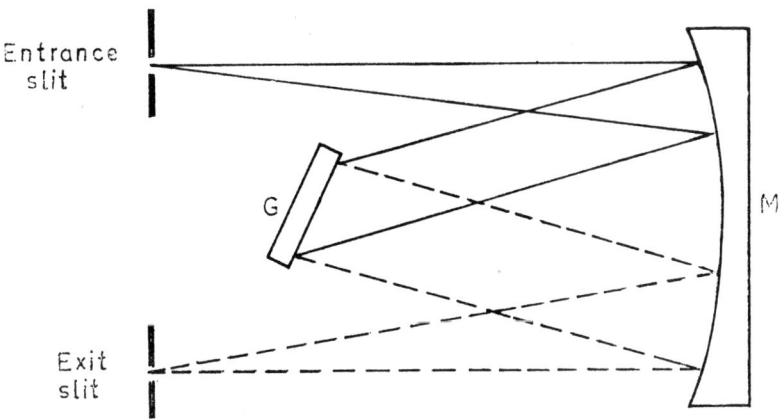

Fig.2.24: Fastie-Ebert mounting

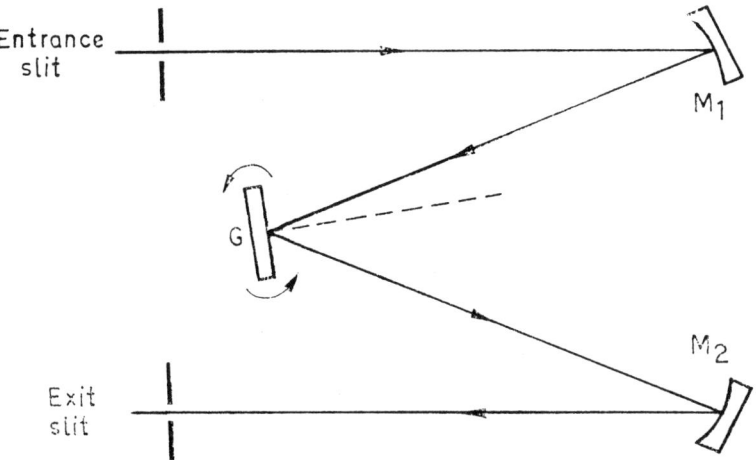

Fig.2.25: Czerny-Turner mounting (Schematic)

are equal i.e. $\alpha = \beta = \psi$. As the grating is rotated through an angle ϕ from its original position, the optical field also moves and in this position, $\alpha = \phi - \psi$ and $\theta = \phi + \psi$. As α and θ correspond to the angles of incidence and diffraction respectively and they are on the same side of grating normal, for a reflection grating we have,

$$m\lambda = d(\sin \alpha + \sin \theta)$$

or $$m\lambda = d\{\sin(\phi - \psi) + \sin(\phi + \psi)\}$$

or $\quad m\lambda = 2d \sin \phi. \cos \psi$

or $\quad \lambda = \dfrac{\{2d \cos\psi\} \sin \phi}{m}$ $\qquad\qquad$ (2.14)

where d is the grating element and m is the order of spectrum. As a result the wavelength, λ, of the light coming on the exit slit is proportional to $\sin \phi$. In fig.2.26, the grating is rotated by a bar fixed normal to the grating with a contact ball that is moved by a plane nut driven by a screw. The bar forms the hypotenuse of the angle ϕ and the motion of nut the ordinate(x), so that a counter geared to the screw will read wavelengths directly. Since $\sin \phi = x/L$, where L is the length of the sine bar, we have, from eqn. (2.14),

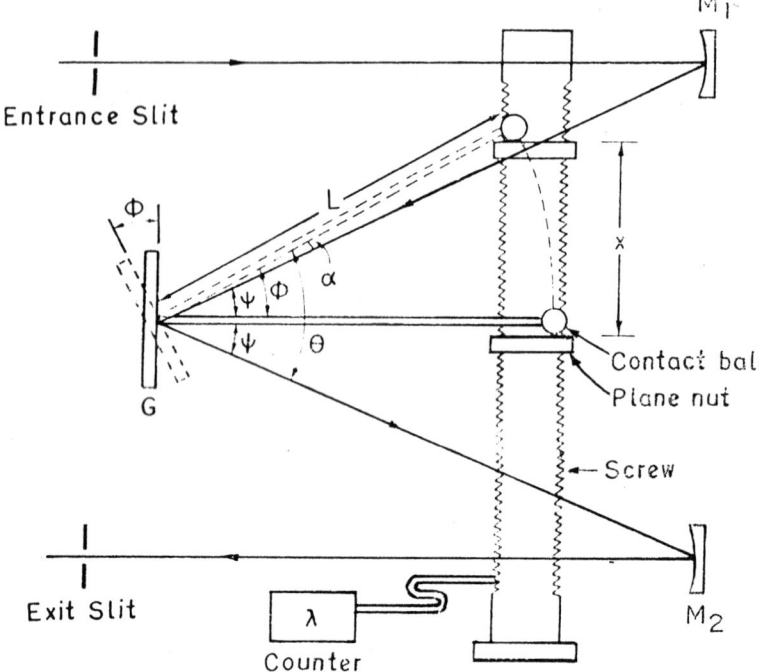

Fig. 2.26: Sine drive mechanism implemented in a Czerny-Turner mount. (M_1 & M_2 are mirrors).

$$\lambda = \frac{(2d \cos \psi)}{m} \times \frac{x}{L} = kx \qquad\qquad (2.15)$$

where, $\quad k = \dfrac{(2d \cos \psi)}{m}, \dfrac{1}{L}$ is a constant.

From eqn (2.15), it follows that λ is directly proportional to the displacement x of the screw. This arrangement is also known as linear wavelength drive. For linear wave-number (cm^{-1}) presentation, secant drives are employed.

Rapid scanning in Littrow type prism monochromators can be achieved by rotating the mirror associated with the Littrow prism. This mirror is generally driven by rotating a cam of appropriate shape to give a linear wavelength or wave number drive.

2.5 DETECTORS

The devices serving as physical detectors of electromagnetic radiation may be divided into two main groups viz(i) thermal detectors and (ii) photon detectors. Their performance, in general may be evaluated in terms of the following parameters.

(i) **Sensitivity :** It is defined as the change in out-put of the device for a given change in input. Sensitivity is measured quantitatively in terms of the spectral responsivity (R_λ) which is the ratio of the rms signal output (voltage or current) to the rms value of the monochromatic input signal power.

(ii) **Noise equivalent power (NEP):** It is defined as the rms value of the sinusoidally modulated monochromatic radiant power incident on the detector, giving rise to a rms signal voltage equivalent to the rms noise voltage (from the detector) in a 1Hz bandwidth. NEP (2 μm, 700) means noise equivalent power at $2\,\mu$m wavelength; 1 Hz electrical bandwidth and 700 Hz chopping rate.

(iii) **Spectral detectivity (D_λ)** It is the reciprocal of $(NEP)_\lambda$.

(iv) **Spectral D-star,** $\{D^* (\lambda, f)\}$, It is the normalized spectral detectivity $D(\lambda)$, taking into account the area and the electrial bandwidth dependance; e.g. $D^*(2\ \mu m,\ 700)$ means D^* at$\lambda = 2\ \mu$m and chopping frequency 700 Hz, unit area and 1 Hz electrical band width.

2.5.1 Thermal detectors

Important detectors in this category are, thermocouples, thermopiles, bolometers and pyroelectric detectors. These detectors are frequently used in infrared spectrometers; however, their response extends over the entire electro-magnetic spectrum and therefore, may be used in other spectrometers also.

In thermocouple or thermopile, incident thermal energy is converted into electrical energy by thermoelectric effect. In practice, the receiver of radiation is made up of a thin blackened flake connected thermally to one of the two junctions of an electric circuit formed by two dissimiliar metals or semiconductors. The other junction is screened and remains cool. The heat absorbed by the receiving junction causes its temperature rise and hence a thermoelectric emf is developed which is proportional to the temperature difference of the two junctions. This can be measured with a potentiometer. A thermocouple employs a single pair of such junctions, whereas a thermopile employs a number of thermocouples that are connected in series. The blackned junctions are clustered to form a target for the incident radiant flux.

The operation of bolometers is based on the measurement of an electrical-characteristic variation caused by the heat absorption in a temperature-sensitive electrical element. In ferroelectric bolometers, the variation affects the value of capacitance. In metal bolometers, called "thermistors" (for thermally sensitive resistors), the variation affects the value of DC resistance, and so on. In all cases, bolometers are not frequency-selective. A typical thermistor bolometer circuit is shown in Fig.2.27. The essential parts of this type of bolometer are two thin and narrow ribbons of suitable material (for example, platinum) forming two arms of a wheatstone bridge. In figure 2.27, R_A and R_C denote the active and compensating thermistors (R_A generally blackened to enhance the response). When R_A is irradiated, its resistance is increased as its temperature increases, whereas R_C (which is shielded from the radiation) remains same. Therefore the balance of the bridge is upset and a voltage appears at the output terminals.

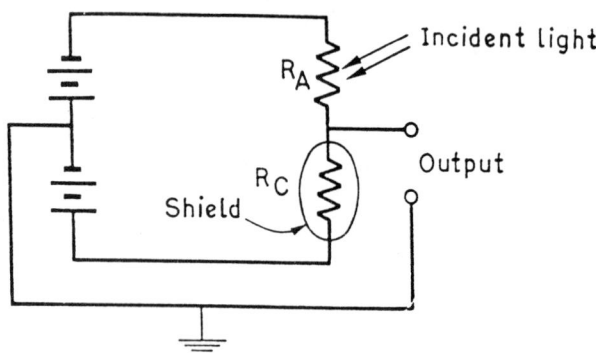

Fig. 2.27 Typical thermistor bolometer

In pyroelectric detectors, a ferroelectric material (e.g. triglycine sulphate; TGS) which exhibits spontanaous polarization (electric charge concentration) and is temperature dependent, is used. In such a device a small capacitor is formed by two metal plates enclosing a TGS crystal as the dieletric. As the incident radiation is absorbed by the dielectric, a voltage appears at the two terminals of the capacitor, which is proportional to the incident radiation.

2.5.2 Photon detectors

These are devices in which the incident photons produce the electrical effects directly. Depending upon the nature of the phenomena occuring on the absorption of photons, they can be classified as, photoemmissive, photoconductive or photovoltaic detectors. In a photoconductive device, the incident radiation causes the change in conductance of the material employed; in a photoemissive device, the incident photons cause the emission of electrons from a sensitized surface and in a photovoltaic device, the radiation incident on the p-n junction produces free electron hole pairs which are separated by the internal electric field to cause voltage (or current). The sensitivity of these detectors is strongly dependent on the wavelength of the incident radiation. This characteristic is limitation for most purposes; however, it can serve a good purpose for filtering out the stray radiation of undesirable wavelengths.

In general, the photoconductive detectors are not suitable for precision measurements because

(i) their response to the incident radiant flux is highly non-linear,

(ii) their response time is long and

(iii) they show fatigue effect.

However, they show larger sensitivity which depends on the size of the cathode and the external voltage applied to the detector.

The photovoltaic detectors are finding increasing application in photometers because they are cheaper, robust and fairly sensitive. Whenever, they are to be used in precision measurements; the spectral response (which varies markedly with wavelength) of these detectors must be kept in mind. The out-put voltage and current are not proportional to the illuminance and are dependent on the resistance in the external circuit. The sensitivity varies from detector to detector and also within the area of an individual detector. The sensitivity is also dependent on the angle of incidence of the radiation .

Although the sensitivity of photoemissive detectors are much smaller than that of the photovoltaic detectors (typically around 10 μA/ lumen as against around 500 μA/lumen), they are the most widely used devices. The reasons for this are that the photoemissive detectors

(i) show nearly linear relationship between the photocurrent and illuminance;

(ii) have very short response time (of the order of 10^{-8} sec);

(iii) are highly stable

(iv) the photocurrent for a given illuminance, is nearly independent of the applied voltage (above a saturation voltage) for vacuum detectors and

(v) the photocurrent can be effectively amplified.

A typical vacuum photodiode and the associated circuit is shown in Fig.2.28. The electrodes are sealed in an evacuated envelope which is either transparent in uv and visible regions or contains an optical window which allows only, the radiation of desired spectral range to pass through it. When operated, the electrode with an emitting surface, (generally called a photocathode) is kept at a negative potential with respect to the anode (i.e. a collector electrode).

At lower positive voltages of the anode, the rate of collecting the photoelectrons increases as the voltage is increased till a saturation in reached. Above the saturation voltages, almost every emitted electron is collected and hence the photo-current becomes independent of the anode voltage above this saturation value. The photodiodes are operated in this saturation region and with this condition, the current varies linearly with the intensity of the incident radiation. The spectral response of these detectors is determined by the material of the electron-emitting layer on the cathode and the material of the window. Longwavelength cut off is largely determined by the emitting layer and the short wavelength cut off is decided by the window material. The usual cathode materials are caesium on antimony and oxygen (Cs-Sb-O), caesium on antimony only (Cs-Sb), caesium on silver oxide (Cs-Ag_2O); potassium on silver oxide (K-Ag_2O), caesium on trialkaly (Cs-Sb-Na-K) and so on.

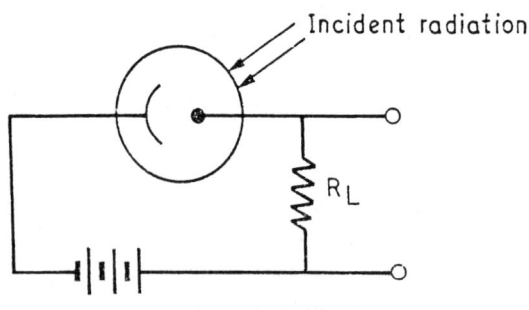

Fig.2.28 Vacuum photodiode

A photomultiplier tube (PMT) is essentially a vacuum photocell made up of a photocathode, a series of secondary electrodes (called dynodes) and an anode. Two designs of the PMT are illustrated in Fig.2.29.

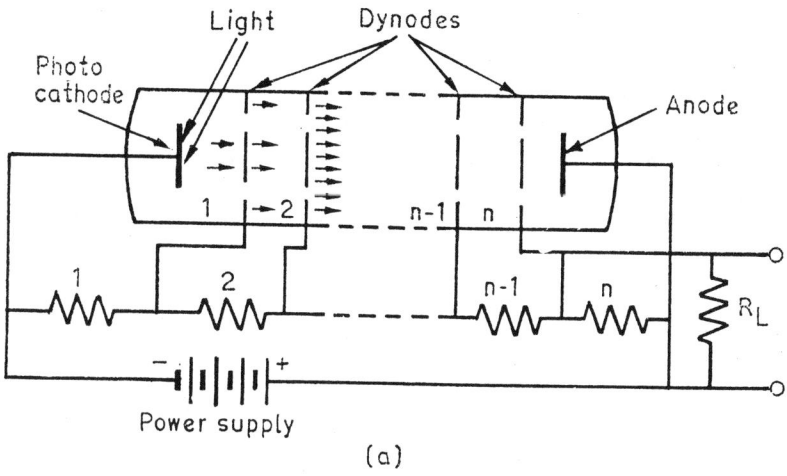

(a)

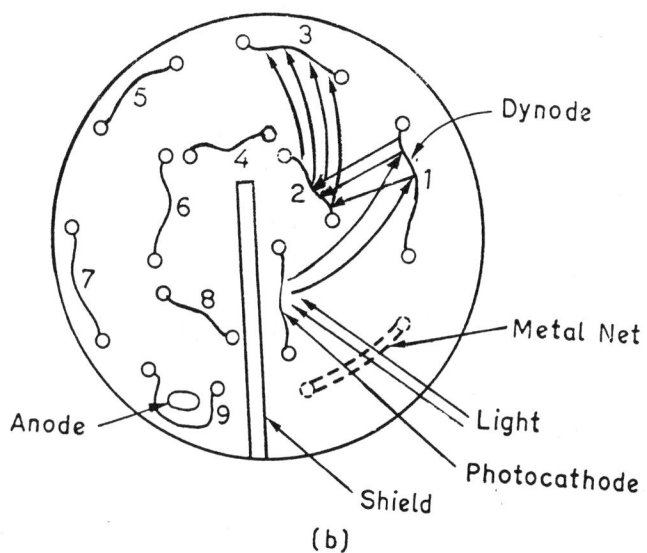

(b)

Fig.2.29 : Schematic design of photomultiplier with
(a) net type dynodes and
(b) central dynode arrangement

Each dynode consists of metal plate coated with a layer, similar to that of a photocathode, so that it readily emits electrons. The arrangement of dynodes is such that each primary photo-electron (emitted by the photocathode) is accelerated towards the first dynode and upon striking it liberates 2 to 5 secondary electrons. These secondary electrons are accelerated towards the next dynode and in this way produce a cascade effect. With 9 to 12 dynodes, an amplification of the order of 10^6 to 10^8 is obtained. Unlike the photodiode, the output current of the PMT increases sharply with increase in the applied voltage. As the total gain of the PMT is extremely sensitive to the changes in the applied voltage, the supply of power must be very well regulated. The spectral response is dependent on the photocathode coating and the window material.

2.6 SOME DESIGNS OF EMISSION SPECTROMETERS

There are two types of emission spectrometers:

(i) simultaneous multielement spectrometers, which are capable of detecting and measuring emission lines for a large number of elements (e.g. 40 to 60) simultaneously; and

(ii) sequential spectrometers, which can measure line intensities corresponding to various elements on a one-by-one basis.

The instruments in the first category can be further divided into three types. They are :

(i) a spectrograph , which employs a photographic film or plate as a detector and integrator

(ii) a photoelectric spectro-meter employing an array of photomultiplier tubes, one for each element, located on the exit curve of the monochromator, and

(iii) an optical multichannel analyzer, employing a vidicon tube as a detector coupled to a computer . In the last one, the scanning over 500 lines or channels, is accomplished electronically rather than optically. However, the resolution of such analyzers is poor as compared to the other two types.

2.6.1 Spectrographs

In these instruments, simultaneous recording of emission lines is achieved by means of a photographic plate or film placed on the exit curve of the monochromator. After exposing the film or plate for a definite time, they are devloped. The developed plate shows a series of black images of the entrance slit. These images correspond to various spectral lines. The location of the lines gives the qualitative information about the elements present in the sample, whereas the intensity of the lines can be related to the concentrations. In the prism type instruments, Littrow mount (Fig.2.13) and constant deviation arrangement (Fig.2.14) are common. In the concave grating type instruments, Eagle mount (Fig.2.21) and Wadsworth mount (Fig.2.22) are generally employed. Concave gratings of length 3 to 4 meters are quite common. For compactness in design, plane grating mounts may be used.

2.6.2 Simultaneous multielement spectrometers

These spectrometers are generally employed for quantitative or routine analysis. They are, therefore, equipped with several photomultiplier tubes located behind the fixed slits along the exit curve of the monochromator. The output of each photomultiplier tube is fed via a suitable integrating circuit to the indicating dials or through an interface to the computer.

Such instruments are highly complex, automated and expensive and permit the detection of only certain elements that are decided at the time of instrument manufacture. However, such spectrometers are ideal for rapid routine analysis, and hence close quality control of the final product becomes possible.

2.6.3 Sequential spectrometers

The cost of multielement spectrometer is very high. Therefore its installation is justified only in situations demanding a large number of routine analyses for a certain set of elements. The plasma sources tend to be sufficiently stable to permit the completion of line-intensity measurements in few seconds. As a result, sequential spectrometers have become a reasonable alternative tò the simultaneous instruments. The cost of sequential instrument is also lower, as this type of design uses a single photomultiplier tube, coupled to an integrator and an amplifier. The readout may be a digital meter or a chart recorder.

2.7 APPLICATIONS

Emission spectrometers are employed for both the qualitative as well as quantitative elemental analysis. The spectrographs are suitable for qualitative or semiquantitative analyses, whereas the simultaneous and sequential spectrometers are useful for quantitative analyses. The sensitivity of measurement depends on the nature and amount of sample, the type of excitation, and the instrument employed . For example, with a d.c. arc excitation, the lower limit of detection may be typically of the order of $10^{-5}\%$ for sample weight of the order of 10^{-3} μg. With plasma jets one can achieve better results.

TUTORIAL-2

2.1 A typical manufacturer's manual describes the monochromator as follows:

Type of mount: Littrow

Aperture ratio : F/8.0

Focal length: 400 mm

Reciprocal linear dispersion: $0.02\mu m/mm$ at the exit slit Type of grating: Reflection grating, 50 mm x 50 mm ruled area, 1200 grooves/mm, blazed for 0.63 μm.

(a) Sketch the optical diagram of the system.

(b) What is the significance of the term "aperture ratio"? Calculate the size of the limiting aperture. How is this size related to the grating size?

(c) Calculate the theoretical half intensity spectral bandpass of this instrument with slits of 50 μm width.

(d) Calculate the resolving power of the grating in the first order.

(e) What is the blaze angle of the grating ?

(f) Predict the spectral range in which the efficiency of the grating is better than 50% in the first order.

SOLUTION

(a) Optical diagram of the system is shown in Fig.2.30 a, and the enlarged cross-sectional view of reflection from a single groove is shown in Fig.2.30 b.

(b) The light gathering power or ability of a monochromator is described quantitatively by the "aperture ratio" or the f-number of the optical system. This is given by

f/- = F/D

where F is the focal length of the collimating device (lens or mirror) and D is the diameter of the circular aperture of the same area as the limiting aperture of the monochromator. The latter is generally decided by the dispersing element (in the present case, by the grating). An increase in D would result in a collection of a larger fraction of radiation.

In the present problem,. since the f-number is 8.0 and F = 400 mm, therefore, the size of the limiting aperture = 400 mm/8 = 50 mm

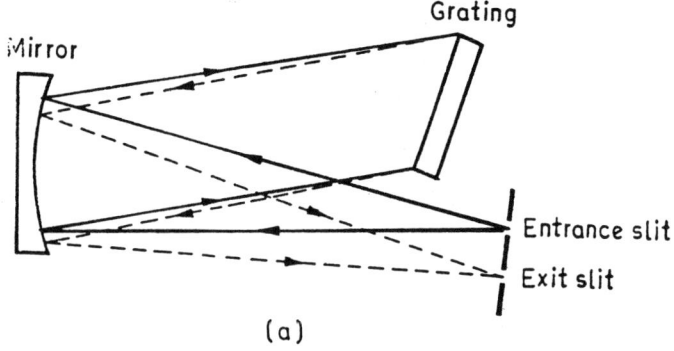

(a)

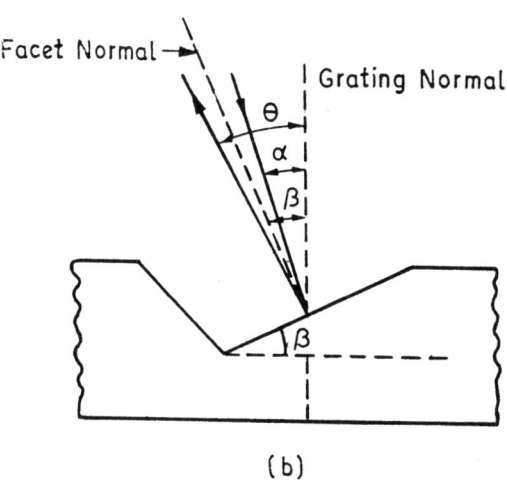

(b)

Fig.2.30

(c) The theoretaical half-intensity spectral band-width or bandpass, S, of a monochromator is given approximately by the relation: $S = W (d\lambda/ dl)$,

where W is the mechanical slit width and $d\lambda/dl$ is the reciprocal linear dispersion.

Since $W = 50 \ \mu m = 50 \times 10^{-3} mm$

and $d\lambda/dl = 0.02 \ \mu m/mm$; we have

$S = (50 \times 10^{-3} mm) \times (0.02 \ \mu m/mm) = 10^{-3} \ \mu m$.

(d) The resolving power, R, of the grating is given by

$R = m N$

where m is the order number of the spectrum and N is number of lines or grooves in the grating.

As there are 1200 grooves per mm, the grating element, d is $d = 10^{-3}/1200m = 0.83 \times 10^{-6}$ m $= 0.83$ μm. And since the width of the grating is 50 mm, the total number of grooves, N would be 1200 x 50 = 60,000.

Therefore, in the first order, (m=1)

R = 60,000 .

(e) In the Littrow mode, the angle of incidence (α) and the angle of diffraction (θ) are nearly equal to the angle of blaze(β) as shown in. Fig.2.30 (b), That is $\alpha \approx \theta \approx \beta$ The fundamental grating equation, therefore, modifies to, taking the blazed wavelength $\lambda = \lambda_B$, $m\lambda_B$ = 2d sin $\theta \approx$ 2d Sin β.

Therefore,

$$\sin \beta = \frac{m\lambda_B}{2d}$$

Here m = 1, $\lambda_B = 0.63$ μm and d = 0.83 μm, and hence

$$\sin \beta = \frac{0.63}{2 \times 0.83} = 0.3795,$$

and $\beta = 22.3°$

(f) The efficiency of the grating falls off on either side of the blazed wavelength, λ_B. In order to predict the spectral range in which the grating efficiency will be better than 50%, usually the following rule of thumb is applied :

$$\left[\lambda_1 = \left(\frac{2}{2m + 1} \right) \lambda_B \right] < \lambda < \left[\lambda_h = \left(\frac{2}{2m - 1} \right) \lambda_B \right]$$

where m is the order number and λ_1 and λ_h are the lower and higher limits respectively of wavelengths.

In the present case, $\lambda_B = 0.63$ μm , and hence in the first order (m = 1),

$$\lambda_1 = \left(\frac{2}{2+1} \right) \lambda_B$$

$$= \frac{2}{3} \times 0.63 \ \mu m$$

$$= 0.42 \ \mu m,$$

and $\quad \lambda_t = \left(\dfrac{2}{2\text{-}1} \right) \lambda_B$

$\qquad = 2 \times 0.63 \,\mu\text{m}$

$\qquad = 1.26 \,\mu\text{m}$

Therefore, the wavelengths (λ) lying in the range $0.42 \,\mu\text{m} < \lambda < 1.26 \,\mu\text{m}$ would be transmitted efficiently in the first order.

2.2 A Czerny - Turner monochromator (optical arrangement shown in Fig. 2.31) is employing a grating that has 1000 grooves/mm.

    (a) What wavelengths will appear at the exit slit in the first three orders if the incident beam makes an angle of 20° with the grating normal?

    (b) If the blaze angle of the grating is 25°, compute the blazed wavelength, λ_B; in the first order.

    Answer: (a) 0.5156, 0.2578 and 0.1718 μm

            (b) 0.8420 μm

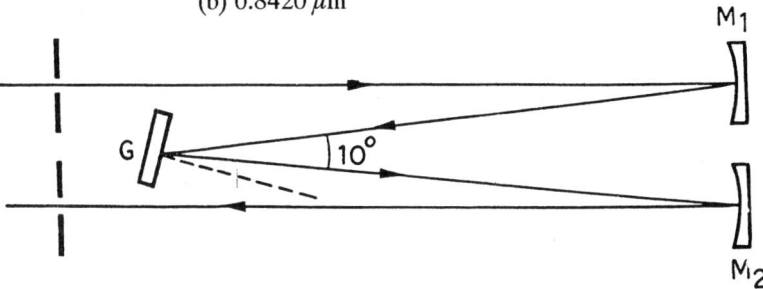

Fig. 2.31

2.3 Compute the maximum resolving power of a grating for a wavelength if the width of the ruled area is W:

    Answer: $R_{max} = 2 \, W/\lambda$

2.4 It is proposed to design a reflection grating for use in the IR. Assume it is to be blazed at $\lambda = 3 \,\mu\text{m}$, have a first order resolving power of 30,000 and a ruled length of 12 cm, what should be (a) the total number of grooves, and (b) the angle of blaze, of the grating.

    (Assume operation in the Littrow mode)

    Answer: (a) 30,000; (b) 22°

2.5 What is the minimum resolving power required to resolve two spectral lines that are separated by 0.2 A° and have a mean wavelength of 5000 A°? Can these lines be resolved with a grating that has 800 grooves/mm and a ruled area of 2 cm x 2 cm?

Answer: 25,000, yes, in the II order

2.6 In a wadsworth mount, shown in Fig.2.32, a cornu prism of apical angle 60° is employed.

   (a) What is the angle of minimum deviation, δ, for Hg-line of wavelength 5461A° ?

   (b) What is the angle of reflection, α, so that the ray incident on the prism and that reflected by the mirror remain parallel?

   (Given: Refractive index, n_λ, of quartz for λ = 5461A° is equal to 1.5442).

   Answer: (a) 41° 6' , (b) 20° 33'

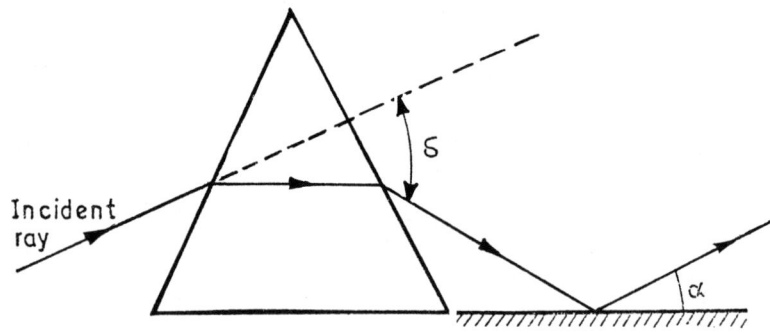

Fig.2-32

Hint (for part b).

It can be shown that, the angle of reflection α is related to the refractive index, n_λ, by the relation:

$\alpha = Sin^{-1} (n_\lambda /.2) - 30°$.

2.7 Explain how the rotating sectors shown in Fig.2.33 may be used as light attenuators? The purpose of such attenuators is to divide the homogeneous illumination at the entrance slit of the spectrograph into a number of sections such that the ratio of the light flux passing through different sections is a known quantity. With such an attenuator the spectral line recorded on a photographic plate consists of several segments having definite intensitiy ratios. Light attenuators are primarily employed for calibrating the emulsion of the photographnic plate or film.

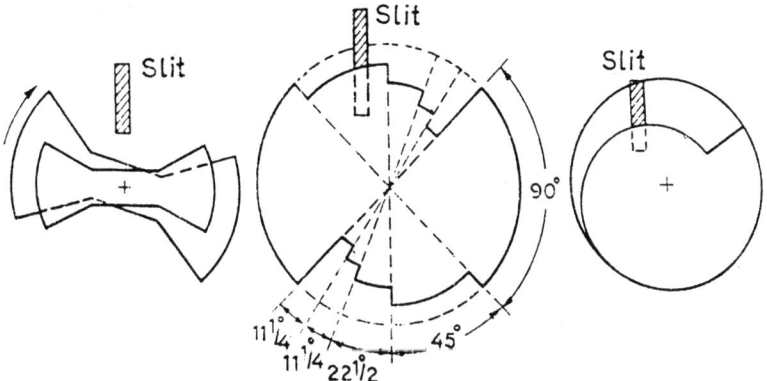

(a) Two-disc
rotating sector

(b) Five-step
rotating sector

(c) Logarithmic
rotating sector

Fig.2.33

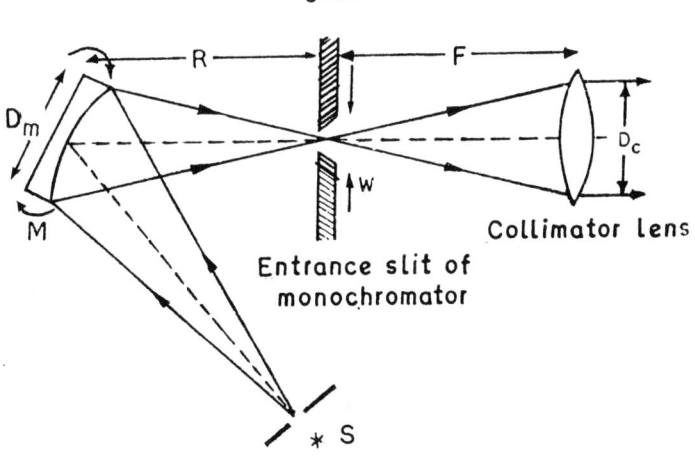

Fig.2.34

2.8(a) A typical arrangement for time resolved illumination of a spectrograph is shown in Fig.2.34. The light source (which is a spark, (S) illuminates the slit of the monochromator via a concave mirror (M) placed on the optical axis of the spectrograph The mirror is driven by a synchronous motor whose axis of rotation is perpendicular to the optical axis of the instrument. What time resolution is obtained if the mechanical slit width W is 0.0314 mm and the distance R between the centres of the mirror and the slit is 50 cm. The mirror is rotated at a frequency of 50 cycles per second.

(b) If the aperture ratio of the collimator lens is 10, what is the diameter (D_m) of the concave mirror M? (R may be taken to be nearly equal to the focal length of the mirror).

Answer: (a) 20 μs, (b) 50 mm

Hint: The image of the source formed by the mirror M moves along the slit with a linear velocity $v = R\omega$, where ω is the angular velocity of the rotating mirror and hence the time of single illumination, Δt (also known as time resolution) would be

$$\Delta t = \frac{W}{v} = \frac{W}{R\omega} = \frac{W}{2\pi Rf}$$

where f is the frequency in cycles, per second. It is essential that the aperture ratios of the mirror M and the collimator lens be nearly equal, (ie. $D_c / F = D_m / R$).

ABSORPTION SPECTROMETERS

3.1 INTRODUCTION

When a beam of electromagnetic radiation passes through a medium, some portion of it is lost by reflection, some is scattered whereas the remaining portion is transmitted but reduced because of the absorption within the medium (see Fig.3.1). For a particular wavelength, the absorption of radiation by the medium is governed by Beer's law which states that

$$\log \frac{P_{o\lambda}}{P_\lambda} = A_\lambda = abc, \tag{3.1}$$

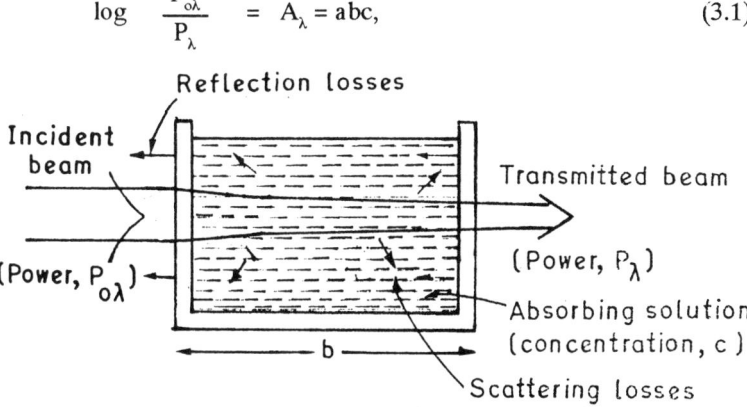

Fig.3.1: The intensity of the incident beam of radiation is reduced on account of the losses due to reflection, scattering and absorption by the medium.

Here $P_{o\lambda}$ and P_λ are the radiant powers of the incident and the transmitted beams. The radiant power is defined as the amount of energy incident per unit area per second. 'A_λ' is the absorbance of the medium, 'a' is a proportionality constant and is known as absorptivity, 'b' is the path length and 'c' is the concentration of the solute in the solvent. The transmittance, T_λ, of the solution is the ratio $p_\lambda / p_{o\lambda}$.

The measurement of absorbance or transmittance as a function of wavelength is called absorptiometry or absorption spectrometry. The instrument that is employed for such purposes is called an absortpion spectrometer. Since the ions, atoms or molecules, when reasonably isolated exhibit an individual set of energy levels, they absorb only those frequencies that correspond to their excitation from one level to another. Thus the measurement of an absortpion spectrum of an unknown substance may help in its identification. The radiant power absorbed by the specimen can be related to the concentration. The sensitivity of this technique is so much that concentrations as small as 0.1 ppm can be detected with ease.

3.2 DESIGN CRITERIA

From the preceding discussion, it follows that the measurement of absorbance or transmittance would require a polychromatic continuous source to be followed by a monochromator. A beam of radiation that is attenuated by the sample may be detected by a suitable detector/transducer coupled to a readout via a signal processor. Thus the modules of an absorption spectrometer may be arranged as shown in Fig.3.2.

How is the static performance of this instrument related to the modular performance?

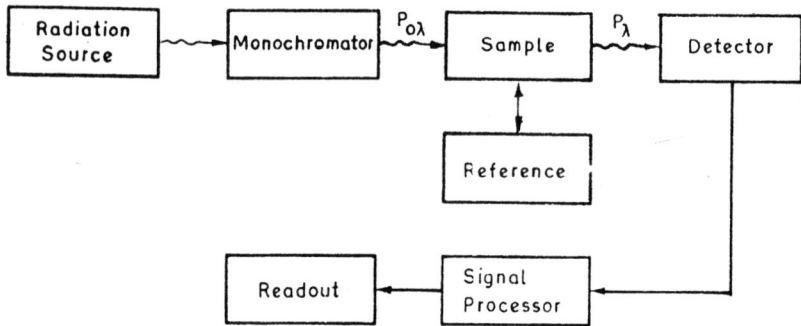

Fig. 3.2 Block diagram of a single beam absorption spectrometer.

The signal, S, developed by the detector/transducer can be related to the modular performance as follows:

$$S = [P (\lambda). M(\theta, \lambda). T(\theta, \lambda)] e^{-abc}. D(\lambda). \qquad (3.2)$$

Various factors in equation (3.2) are identified as follows.

$P(\lambda)$ is the power radiated by the source as a function of wavelength, λ; $M(\theta, \lambda)$ is a factor that governs the solid angle (θ) as seen by the monochromator and its transmittance with wavelength (λ); $T(\theta, \lambda)$ is the transmittance of the reference or blank. Thus the three terms combined together inside the square brackets simply give the power transmitted by the reference or the blank. This can be expressed by $P_{o\lambda}$ of equation (3.1). Substituting for the terms in square brackets in (3.2), we get

$$S = P_{o\lambda} \, e^{-abc} \, D(\lambda) \qquad (3.3)$$

As $P_{o\lambda}$ is the power incident on the sample, according to Beer's law, the transmitted power P_λ will be equal to $P_{o\lambda} \, e^{-abc}$. Finally, the function $D(\lambda)$ represents a factor that gives the response of the detector as a function of wavelength, λ.

It is obvious from equation (3.2) that for obtaining a larger signal, the performance of each module must be optimized. In other words, the modules must be selected according to the spectral range of interest (e.g. UV/VIS/IR) and must be arranged in such a manner so as to avoid losses due to reflection, scattering and absorption and also avoid stray radiation. Thus the source should have some area of uniform intensity. Only this area should be imaged on the entrance slit of the monochromator. The entrance aperture can then be imaged and reimaged employing front-surface off-axis mirrors and/or corrected lenses. It should be noted that the beam cross-section be as small as possible at the site of the sample, so that, if the occasion arises, microsamples can also be analysed with no change in optical design.

As a rule of thumb, the monochromator precedes the sample in UV-VIS instruments, whereas it follows the sample in IR spectrometers. This is done to minimize the stray radiation coming inside the photometric system.

3.3 COMPONENTS

3.3.1 Radiation sources

The success of absorption measurements depends considerably on the proper choice of the radiation source. The latter are classified as continuous or discrete. The continuous sources are those whose spectrum extends over a considerably large spectral region. Continuous spectra are, generally, emitted by black body radiators, incandescent solids, incandescent liquids and certain high pressure discharge tubes. The spectrum of a discrete source is characterised by discrete or sharp lines. Such a spectrum is emitted by an atom or molecule in a gaseous or vapour state which acquires excess energy in some manner to be excited to higher levels; subsequent de-excitation giving rise to the radiation. Electric arcs, sparks and low-pressure ionic discharges are examples of discrete sources. The discrete sources are, generally not employed in absorption measurements and hence we shall not be discussing them here. In the following paragraphs, we discuss the continuous sources only.

Incandescent non-gaseous sources

An incandescent body may be defined as one whose radiation is due to its temperature. This radiation is characterised by a broad continuous spectrum and closely resembles that of a blackbody. The total amount of energy radiated by a black body(or approximately that by a hot body) is proportional to the fourth power of its temperature (in kelvin). The relationship, known as Stefan-Boltzmann law is as follows:

$$P = \sigma \, T^4 \; (W/m^2) \tag{3.4}$$

where P is the total power radiated per unit area of the radiating surface, T is the absolute temperature of the body and σ is the Stefan-Boltzmann constant and is given by

$$\sigma = \frac{2\pi^5 k^4}{15 h^3 \, C^2} = 5.6697 \times 10^{-8} (W/m^2 \, K^4) \tag{3.4a}$$

Here k is the Boltzmann's constant, h is the Planck's constant and C is the speed of light.

Relation (3.4) does not give any information about the distribution of energy in the spectrum. This is provided by Planck's distribution law, which states that the monochromatic emissive power of the black body is inversely proportional to the fifth power of the wavelength. Mathematically, it is expressed as

$$P_\lambda = \frac{C_1}{\lambda^5} \left\{ \frac{d\lambda}{\{\exp(C_2/\lambda T)-1\}} \right\} \tag{3.5}$$

where P_λ is the power radiated per unit area for radiation whose wavelength lies between λ and $\lambda + d\lambda$; the first radiation constant, $C_1 = 2\pi h C^2 = 3.7415 \times 10^{-16}$ (W.m^2) and the second radiation constant, $C_2 = h \, C/k = 1.43879 \times 10^{-2}$ (mk).

Fig.3.3 depicts the plots of eqn (3.5) for an ideal radiator (i.e., a perfect black body) at various temperatures. The area under each curve gives the total energy radiated. This area and hence the energy radiated at a particular temperature may be computed by integrating eqn (3.5). from $\lambda = 0$ to $\lambda = \infty$; the result being eqn (3.4). The curves shown in Fig.3.3 are representative of all incandescent bodies. With increase in temperature, the shape of the curve remains the same but the ordinate increases and the maximum in the curve shifts towards shorter wavelength side. This shift may be calculated employing Wien's displacement law, which states that the wavelength of maximum emission λ_m, of a black body radiator is inversely proportional to its absolute temperature (T). Mathematically,

$$\lambda_m = \frac{B}{T} \qquad\qquad (3.6)$$

where B is a constant. If λ_m is measured in μm, B equals 2898.

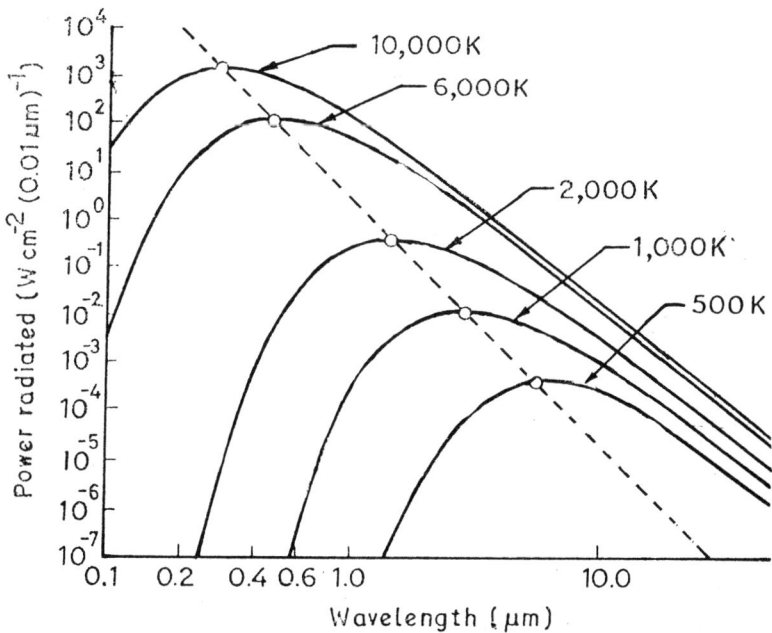

Fig.3.3 : Spectral distribution of a black body radiator at various
temperatures. The peak values of radiated energy at different
temperatures are shown by circles on the dashed curve, which
obeys eqn.(3.6).

From the above discussion, it is evident that in order to obtain a high
output of visible or ultraviolet radiation from an incandescent body, a high
temperature would be required. A convenient approach for making use of the
radiation from incandescent bodies is the incandescent lamp, so let us discuss
a few specific-lamps of this type.

Nernst glower : It is generally designed in the form of a long thin cylindrical
rod or tube from refractory materials such a zirconia, yttria or thoria. Platinum
leads are provided at the ends of the rod or the tube to supply power to the
glower.

Since the glower rod is a semiconductor it must (a) be preheated to start it and (b) be used with a ballast in the circuit to prevent burnhout. The useful spectral range of the glower is from approx. 0.40 to about 15 μm. Its life depends on the operating temperature and care in handling. Because of relatively low cost and its shape, it is very well suited for illuminating the spectrometer slits.

Globar: It is also constructed in the form of a rod but from bonded silicon carbide. Metallic caps are provided to serve as electrodes. The current through the globar heats the element , i.e., the rod in the color temperature range from about 1200 - 1500 K. An extra facility is needed to cool the container of the rod and this makes it little more expensive than the Nernst glower. The useful spectral range of the globar is from 1 to about 40 μm. IR spectrophotometers, normally employ this source.

Tungsten filament lamps

A conventional tungsten filament lamp consists of a filament of fine tungsten wire mounted inside a glass envelope which is either evacuated or filled with an inert gas. The filaments have different designs (i.e., different sizes and shapes) depending on the use to which the lamps are designed. In general, the filaments are closely wound coils or coiled coils. The efficiency of lamps increases with increasing the temperature of the filament. The operating color temperatures range from 2400K to about 3500K and with glass envelope the spectral range from 0.35 to about 2.5 μm is obtained.

In some lamps, known as tungsten - halogen lamps, the blackening of the envelope is eliminated by the internal chemical cycle that returns the filament most tungsten that sublimes. A halogen (e.g. iodine) vapour combines with the tungsten which is deposited on the envelope wall by the evaporation of the filament. It forms a volatile compound (e.g. WI_2) which decomposes on the surface of the hot filament, thus redepositing tungsten on it and leaving the halogen free to repeat the cycle. Such lamps are more efficient and their life also longer and hence they are normally used in uv/vis spectrophotometers. With tungsten iodine lamp at 3500 K and with quartz envelope an useful spectral range from 0.2 to 0.8 μm is easily obtained.

High pressure sources

One of the important properties of the high pressure arcs is the high temperature (which ranges from 5500 to 8000K) of the gas and the thermal equilibrium between the gas and the electrons. These sources are characterised by an intense small diameter arcs operating at relatively large current density and high voltage. As the potential gradient in the arc increases with increase in pressure, the high pressure arcs are shorter than low-pressure arcs at the same voltage. The spectrum of these sources show a broadening of lines with increasing pressure and a continuous background which also increases with the pressure (probably due to greater molecular interaction).

Compact source arcs

Compact sources have been classified into three types depending on the gas or vapour contained in the arc envelope. They are :

(i) mercury arc,

(ii) xenon arc, and

(iii) mercury xenon arc.

Although they are similar in many aspects, the presence of different gases contributes to significant differences in the performance.

The common features are as follows. These sources are designed for a.c. or d.c. operation. For a.c. operation, both the electrodes are, generally, similar, whereas for d.c. operation, the anode is much larger than the cathode. The sources are operated best with the arc axis vertical or nearly so. A source operated on a.c. has nearly uniform luminance gradient along the length of the arc, whereas that operated on d.c. show highest luminance at the "hot spot" near the cathode and decreases rapidly in the direction of the anode. In both the cases, arcs give highest radiant intensity and the luminous intensity in a direction perpendicular to the arc axis. At angles above and below this axis, both these intensities fall off gradually.

The spectral distribution of the three types of sources differ markedly. The spectrum of mercury lamp consists of characteristic mercury lines and a relatively weak continuous background superposed on it. As noted earlier, the intensity of the background increases with increasing pressure. The major portion of the radiation is at visible rather than at u.v. or infrared, however, the spectral distribution can be modified by changing the pressure or by incorporating other additives. Apparent color temperatures are in the range of 7000 to 10,000K.

The spectrum of xenon-arc is almost continuous in the entire optical region with some strong lines in the 0.8 to 1.0 μm band. Changes in the operating pressures do not result in any significant changes in the spectrum. The colour of radiation from xenon lamp resembles that of the sunlight with an apparent color temperature of about 6000K.

The spectrum of mercury-xenon arc is almost similar to that of mercury arc except that xenon contributes to the continuity of the spectrum and some of lines particularly those in the 0.8 to 1.0 μm band.

All these sources require very high ignition voltages for starting purposes. Specifically, a mercury lamp, at room temperature, requires several hundred volts across its terminals for starting and requires several minutes for stabilization (i.e. warm up). In some lamps, there is provision of a third or "starting" electrode to which a high voltage ignition pulse is applied. This feature makes the ignition device less expensive as well as lighter. Mercury-xenon and xenon lamps, at room temperature, also require a high voltage (30-50 KV) but high frequency ignition pulse for starting. Both these lamps reach quickly to the stabilized condition.

A stable operation of all the compact-source arc lamps requires the use of either a ballast in series with a lamp on voltage regulated circuit or a highly stabilized power supply.

Other high-pressure sources

There are a number of other high-pressure arc sources, a few of which are summarised as follows.

'Uviarc', (a trade name of General Electric) is an intermediate - pressure mercury arc lamp. It is an efficient source of ultra-violet radiation, though effective spectrum typically ranges from 0.22 to about 0.6 μm. 'Lucalox' (a trade name of General Electric) is a high pressure sodium discharge and high temperature withstanding ceramic lamp. Further, there are 'multi-vapor arc' lamps, which radiate chiefly in the visible region. They contain argon and mercury to provide the starting action, afterwhich sodium iodide, thallim iodide and indium iodide vaporise and dissociate to yield the bulk of the lamp radiation.

'Capillary mercury arc' lamps are designed in the form of small diameter, heavy wall quartz tubes which are enclosed in the outer jacket of quartz to pass u.v. radiation or that the pyrex to prevent u.v. radiation. The outer jacket serves as a filter, heat insulator and also as a shield in the case of any eventuality. The operating pressures range from 2 atmospheres for air-cooled sources to more than 100 atmospheres for water cooled systems. These are useful sources of u.v. as well as visible radiation.

A hydrogen or deuterium arc lamp gives a strong continuous spectrum in the u.v. (typically from 0.2 to 0.4 μm). With suitable windows or slight modification in the design of the lamp, it can yield radiation below 0.2. μm. This lamp is normally used in uv/vis spectrophotometers.

3.3.2 Optical filters

A number of situations are encountered during absorption measurements where the separation of wavelengths is required but at the same time, a high degree of spectral purity achievable with a monochromator is not needed. For instance, it may be desirable to isolate a particular band of wavelength from the working radiation or it may be enough to eliminate all the radiation below or above a given wavelength. Optical filters are designed to serve this purpose.

In general, an optical filter may be defined as a device or material which changes selectively or non-selectively the spectral intensity-distribution or the state of polarization of the electromagnetic radiation incident upon it. According to the nature of mechanism responsible for filter action, the filters have been classified into the following groups; viz. (i) absorption filters; (ii) interference filters, (iii) reflection filters, (iv) christiansen filters (based on scattering of light) and (v) polarization filters.

Absorption filters

The mechanism of absorption is responsible for the filtering action of the absorption filters and is governed by the Beer-Bouguer-lambert law.

These filters are made of glass, gelatin or liquids in which the coloring agents are dissolved or suspended. Although,there exists a number of solids, liquids and gases which possess absorption properties in the desired spectral regions but most of them are not suitable as filter materials because of the instability of their optical behaviour.

Color glass filters

Color glass filters are composed of base or host glasses e.g. sodium, potassium, phosphate, borate or silicate glasses. The colorants in the form of metallic or non-metallic ions, in solution, are added in these glasses. The spectral characteristics of these filters are, therefore, dependent on both the type of glass as well as the colorant used. In some cases, the spectral properties also depend on the heat treatment given to these glasses.

Hundreds of colored glass filters (both the bandpass as well as cut off type) of spectral transmittance in different spectral regions are available.

Gelatin filters

They are made by mixing organic dyes in gelatin solution and coating the mixture on glass plates. When the coatings get dried, the gelatin filters are removed from the plates. They are then, either lacquered or cemented between the glass plates. Such filters are generally used for cutoff purposes.

They are cheaper but become fragile and relatively unstable at temperatures above 50°C in humid atmosphere or in some cases, under the action of light.

Liquid filters

There exists a number of liquids and solutions which show a wide variety of spectral transmittance curves, but many of them lack adequate stability. Their use is mostly confined to specific applications. Generally, glass or gelatin filters are preferred as they are more convenient to handle.

Absorption filters for special applications

A number of absorption filters and their combinations are available for specific use. For example, heat absorbing glasses may be used to absorb the infra-red part of the electromagnetic spectrum and transmit the u.v. and visible wavelengths; color temperature conversion filters may be employed to alter, by selective transmission, the spectral distribution of an incident black body radiation at temperature T_1 to a distribution of radiation at another temperature, T_2-Special glasses, such as, didymium or holmium glasses, which show distinct maxima and minima in their transmittance curves, may be used conveniently for initial calibration of spectrophotometers or for periodical checks of the wavelength scales. "Narrow-bandpass filters" may be employed to isolate particular spectral lines of radiation sources. Such filters are usually combinations of absorption and interference filters and are designed to select a narrow region of the spectrum within which the incident radiation is transmitted, whereas other regions of the spectrum of that radiation are absorbed or eliminated. Similarly, "cut off filters" may be effectively used to reduce the stray light in monochromators. These filters are designed such that they have nearly zero transmittancae at shorter or longer wavelengths and high constant transmittance at longer or shorter wavelengths. "Neutral density filters" may be used to reduce the intensity of the incident radiation uniformly throughout an extended part of the spectrum. Fig.3.4 shows the transmittance curves of the last three types of filters.

Interference filters

The isolation of narrow spectral bands, to a great accuracy, can be accompalished by the mechanism of interference. A Fabry-Perot interference filter, which is based on the principle of Fabry-Perot interferometer, consists of two transparent but partially reflecting plates, separated by a transparent dielectric spacer (as shown in Fig.3.5). A portion of the entering beam (a) is transmitted through

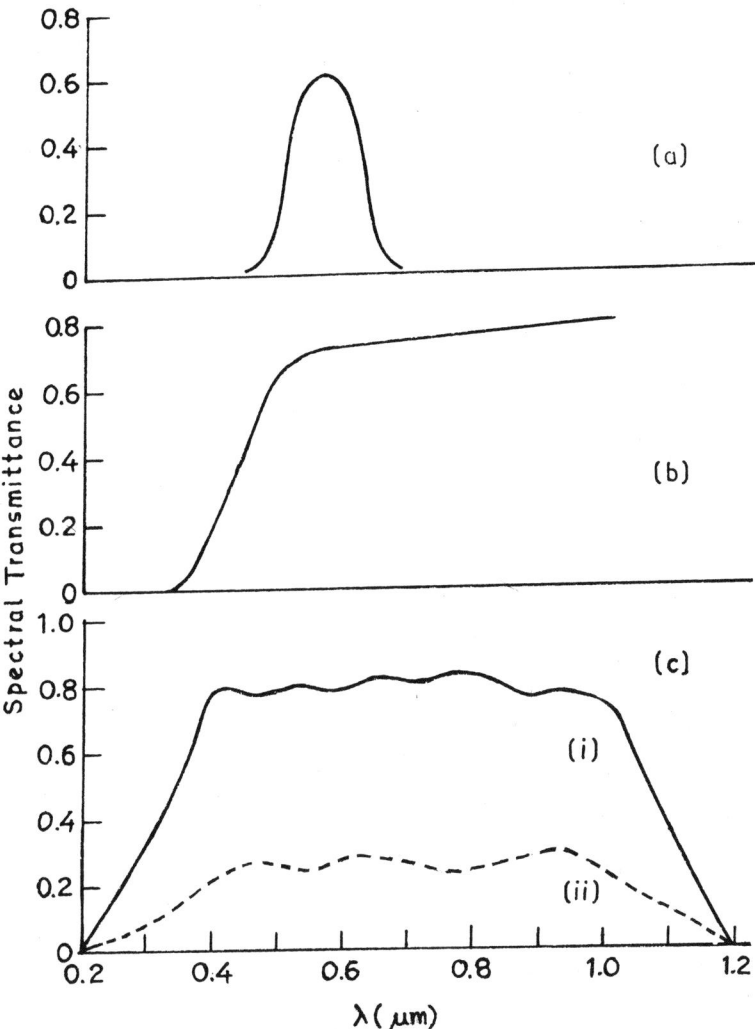

Fig.3.4 : Spectral transmittance curves (schematic) of
(a) narrow band pass filter,
(b) cut-off filter (short wave length cut off);
(c) Neutral density filter (curve ii); curve(i) shows the transmittance of a typical glass plate.

surface (2) as beam (b) while some portion is reflected as beam (c) This beam (c) again suffers a reflection at surface (1) and emerges through surface (2) as beam (d). This process of reflection and transmission continues till the other end of the plates is reached. Considering only the first two emerging beams, (b) and (d), we find that beam (d) traverses distance inside the dielectric material twice more than that traversed by beam (b). If the incident radiation is at right angles to the filter, and if the thickness of the spacer is t, then these two beams will be in phase and will reinforce each other at wavelengths which are integral multiple of 2t (a condition for constructive interference).

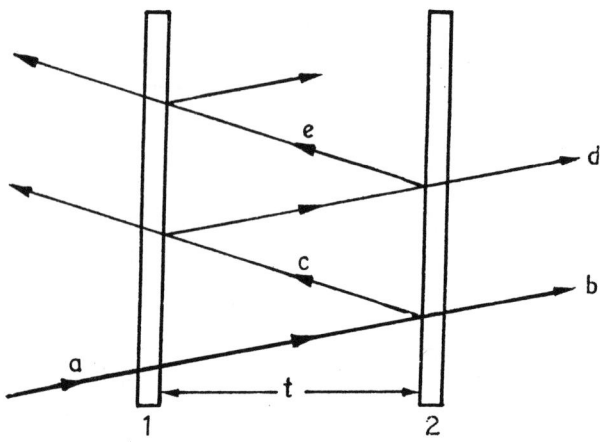

Fig. 3.5 : Schematic diagram of a Fabry-Perot interference filter. For clarity, the distance between the two plates is shown excessively large. Normally, t is of the order of few μm.

Mathematically

$$\frac{m\lambda}{n} = 2t \text{ (for normal incidence)} \tag{3.7}$$

where m is an integer and is known as the order of interference, λ is the wavelength of the incident radiation (in vacuum) and n is the refractive index of the spacer. At other wavelengths, which do not satisfy the above conditions, destructive interference will occur and hence the intensity of the radiation at those wavelengths will be reduced. A filter of thickness corresponding to m = 1(i.e, t = $\lambda/2n$), is called a first order filter, a filter of thickness corresponding to m = 2 (i.e. t = λ/n) is called a second order filter and so on.

In Febry Perot type of filters the transmittance is limited by the absorption of the metallic reflecting films that are deposited on the plates. The modern trend, therefore, is to replace the metallic reflecting films by all-dielectric multilayer stacks. By this method, narrow bandpass filters of extremely high transmittance and with controlled degree of accuracy can be obtained. Such filters are prepared by successive deposition (on a suitable substrate) of 5 to 25 layers of high and low refractive index dielectric materials in alternate layers. The interfaces of high and low refractive index materials serve as partial reflectors and the interference effects described above occur at each interface.By a suitable choice of dielectric materials, number of layers, their thicknesses, substrate materials etc., the filters can be designed for a wide range of wavelengths, half-widths and with desired transmittance characteristics.

Reflection filters

A majority of reflection filters are produced by evaporating metals and /or dielectrics onto glass or quartz substrates. Silver (Ag) shows highest reflectance in visible and infrared and hence it is used for interferometer mirrors and interference filters. A number of metal dielectric reflectors, which reflect highly in one spectral region but not in other regions have been developed for special applications. Such filters may be used to remove stray visible light from infrared and ultraviolet optical systems. Metals and metal-dielectric coatings can also be used in multiple reflection arrangement to make low-wavelength cut-off filters.

A portion of the incident energy reflected by this type of filter, is determined by the Fresnel reflection coefficient of the coating materials; the remaining portion of the incident energy is absorbed by the material. In some cases, where the protecting layers (of either SiO_2 or MgF_2) are used, the reflectance of the filter, will also depend on the spectral properties of these layers.

Christiansen filters

In these filters, the phenomenon of optical scattering is used to remove the unwanted wavelengths from the beam of incident radiation. This type of filter consists of a transparent cell containing a closely packed coarse powder of homogeneous, isotropic, transparent material suspended in a liquid. As the powder and the liquid have different dispersions, they have a common refractive index only at one wavelength, say λc. Therefore a narrow band peaking at λc, from the incident radiation, will be transmitted by the cell without deviation, whereas all other wavelengths present in the incident radiation will be scattered. With proper optical design, the narrow band centred around λc can be isolated and the cell can serve as an efficient narrow bandpass filter. By a suitable choice of the powder and the liquid, christiansen filters for the ultraviolet, visible as well as infrared regions may be obtained.

Some specific combinations are : quartz suspended in a mixture of benzene and alcohol ($\lambda c = 0.3$ to $0.4\,\mu m$); glass suspended in ethyl salicylate ($\lambda c = 0.43$ to $0.57\,\mu m$); NaCl suspended in carbon-di-sulphide ($\lambda c = 5.5\,\mu m$) and MgO suspended in $CC1_4$ ($\lambda c = 9.0\,\mu m$).

Polarization filters

The principal mechanism for filter action in these filters is polarization. They are, therefore, composed of birefringent crystal plates, fractional waveplates or sheet polarizers. The simplest type of polarization filter consists of a series of N birefringent crystal plates of thickness, $t_r = 2^{r-1}\, t_1$ (where r = 1,2,.....N);separated by N + 1 polarizers. The optic axes of all the crystal plates are made parallel and the planes of polarizations of the polarizers are inclined at an angle of 45° to them. When a radiation passes through a single birefringent plate of thickness t, a retardation $\dfrac{t\,(n_e - n_o)}{\lambda}$, is introduced between the components of the beam polarized parallel to the fast and slow axes of the crystal.

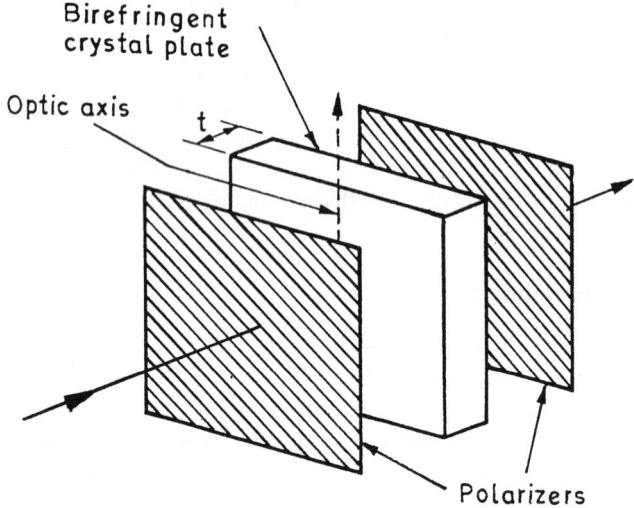

Fig.3.6 : Schematic diagram of a polarization filter.

Here n_e and n_o are, respectively, the extraordinary and ordinary refractive indices of the birefringent plate and λ is the wavelength of the incident radiation. If such a plate is put between parallel polarizers (as shown in Fig.3.6) its spectral transmittance is given by

$$T_\lambda = \cos^2 \left\{ \frac{\pi t(n_e - n_o)}{\lambda} \right\} \qquad (3.8)$$

and the halfwidth ($\Delta\lambda$) of the bandpass is given by

$$\Delta\lambda = \frac{0.5 \lambda^2_m}{t (n_e - n_o)} \qquad (3.9)$$

where λ_m is the peak wavelength of the bandpass.

The spectral band transmitted by a polarization filter is determined largely by the thickness of the thickest plate in the assembly and the spacing between the bands is determined by the spacing between the maxima of transmittance of the thinnest plate. These filters are used whenever very narrow transmission band filtering with tuning capability is required because with such filters it is possible to shift the position of the bandpass in a controlled manner.

3.3.3 Beam splitters

The function of a beam splitter is to divide a beam of optical radiation into two beams of approximately equal relative spectral composition but propagating in two different directions, generally, at right angles to each other. These splitters are designed in two forms, which can achieve

(i)   beam splitting in space, and

(ii) beam splitting in time.

Beam splitting, in space, can be done by four different devices as shown in Fig.3.7. The simplest of these splitters consists of a plane transparent plate on which is evaporated a partially reflecting coating (Fig.3.7a). The second type of splitter, known as cemented beam splitter (shown in Fig.3.7b), consists of a reflection coating sandwiched between two identical plane parallel transparent plates. It is used, where the two derived beams are required to traverse identical paths. A beam splitter cube (shown in Fig.3.7c) is designed to avoid the lateral displacement of the transmitted beam. In a pellicle beam splitter (Fig.3.7d), the reflection coating is deposited onto a pellicle (thin substrate of approximately 1 μm thickness). The last one may introduce an interference pattern into the spectral reflectance or transmittance pattern, particularly, if the beam is monochromatic. In all these splitters, the reflection and transmission coefficients depend on the state of polarization of the incident radiation.

In all the above splitters, each of the derived beams has half the energy of the original beam. This may be a disadvantage, if low signal intensities are to be handled. Alternatively, the incident beam may be intersected by a rotating sector mirror as shown in Fig.3.8. The sector mirror alternately transmits the beam (through an open sector) in one direction and reflects it (via the mirror sector) in the other direction.

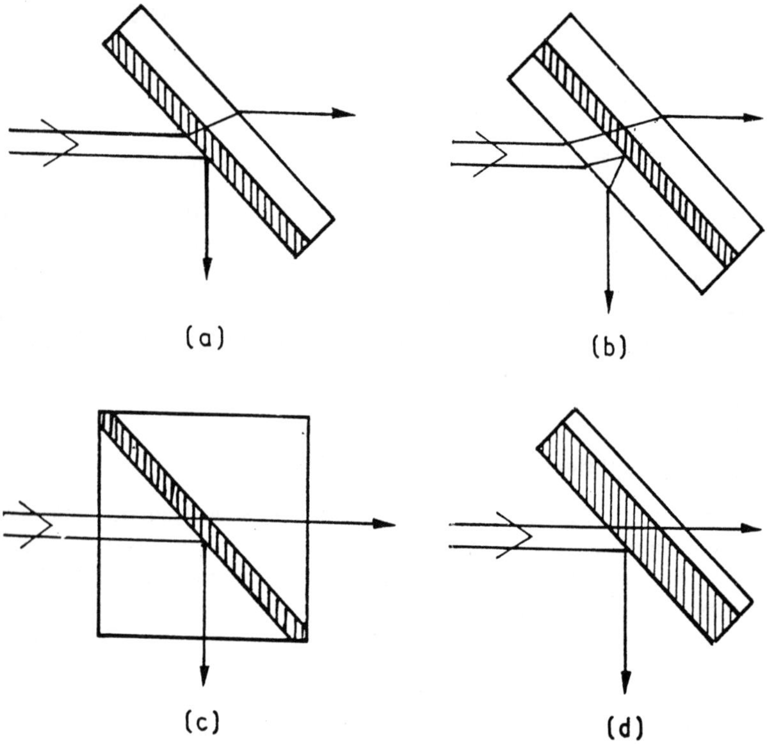

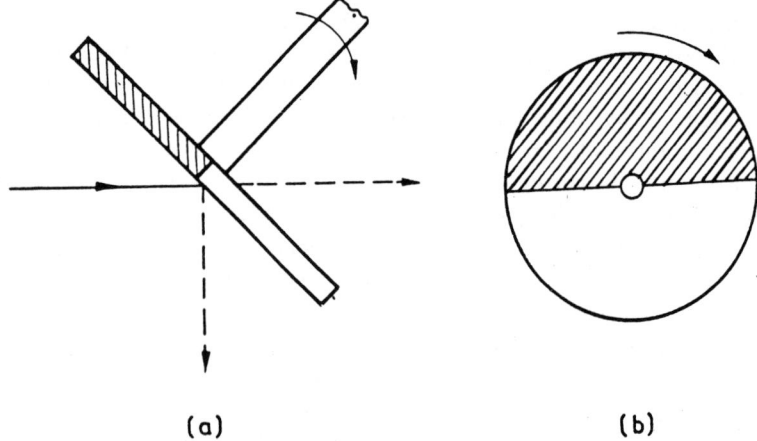

Fig.3.7 : Beam splitting in space, through (a) single plate (b) two cemented plates, (c) cube, and (d) pellicle

(a)

(b)

Fig.3.8 (a) Beam splitting, in time, using a rotating sector mirror;
(b) Sector mirror; shaded area reflects and unshaded area
transmits the incident beam.

The intensity of both the derived beams is the same as that of the original beam. In this time sharing procedure, the splitting is combined with modulation. The latter may be advantageously used in improving the S/N ratio.

3.3.4 Monochromators

Except for the filter photometers, all the absorption spectrometers employ monochromators for dispersing the radiation from the source. Their design aspects are the same as discussed in sec 2.4. Of course, the choice of optical materials for the components depends largely on the spectral range of interest. For example, the materials of the prism or optical window or lens for uv, visible and i.r. regions are normally quartz, glass and NaCl or KCl crystal respectively.

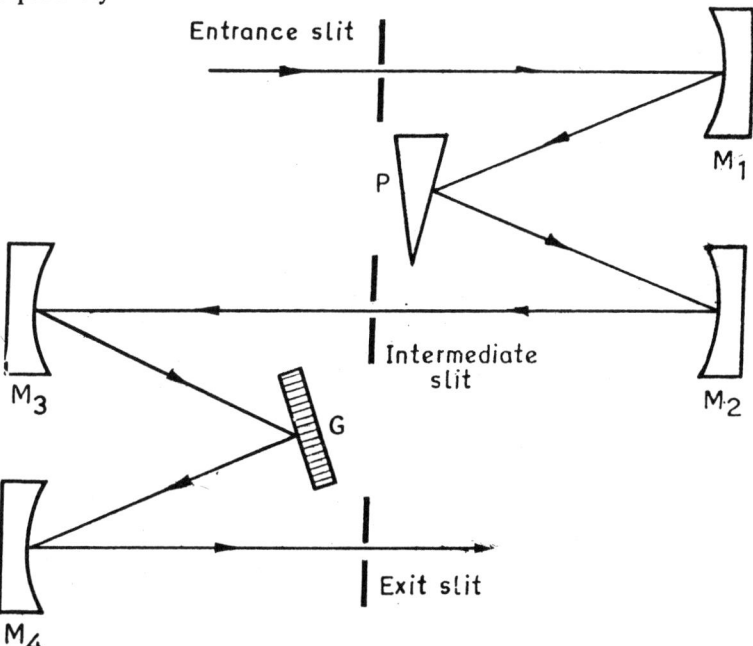

Fig.3.9 : A schematic diagram of a double monochromator. P is littrow prism, G is grating, M_1, M_2, M_3, and M_4 are mirrors

In a monochromator, if the grating is used as a dispersion device, the problem of overlapping orders is always encountered. A widely used solution to overcome this problem is to use an order-sorting device in conjunction with a grating. For example, a littrow prism in series with a grating may be used to form, what is known as a double monochromator, as shown in Fig.3.9. Such a double monochromator is an integral part of all the Cary model spectrophotometers. Here the spectrum dispersed by the prism is focussed onto the intermediate slit which allows only a narrow band of wavelengths to pass through it. The latter is further dispersed by the grating. In this way, the resolution of the monochromator is doubled and at the sametime, the stray radiation because of overlapping orders is also minimized. In this configuration, prism P is known as an order sorter prism. An optical filter at the exit slit of a single monochromator can also be used for order sorting.

3.3.5 Detectors

In UV/VIS spectrometers, photocells and photomultiplier tubes are generally employed for the detection of radiation, whereas, in IR-instruments, thermocouples, photoconductive cells, Golay's detector or bolometers are used. The detailed discussion of detectors may be seen in sec.2.5.

3.4 SOME DESIGNS OF ABSORPTION SPECTROMETERS

Absorption spectrometers may be broadly divided into three categories. They are :

     (i)   filter photometers,

     (ii)   single channel spectrophotometers and

     (iii)   double channel spectrophotometers.

     The first one uses a set of optical filters for wavelength isolation whereas the other two employ monochromators.

3.4.1 Filter photometers

A filter photometer is a simple and relatively inexpensive tool for performing absorption analysis. Favourable features of such photometers are convenience, ease of maintenance and ruggedness. These instruments should be preferred for routine quantitative analysis where high spectral purity is not an important factor. Fig.3.10 depicts schematically two such photometers. The instrument shown in Fig.3.10 (a) is a single beam device consisting of a tungsten filament lamp, a collimating lens, a filter and a photovoltaic detector. The readout is generally a microammeter. 100% transmittance setting is obtained through a reference cell and a variable diaphragm. Fig.3.10 (b) represents schematically a double beam, electrical null-type photometer.

     Here the collimated optical beam is split into two parts by the mirror. One part passes through the sample cell and falls onto a measuring detector (generally a photovoltaic cell) and the second part passes through the reference cell and is detected by a similar reference detector. The currents from the two detectors are passed through the variable resistors R_1 and R_2 and a sensitive galvanometer is used as a null indicator. R_2 is employed for 100% T adjustment while R_1 is calibrated in terms of percentage transmittance (%T). The majority of commercial photometers employ the double beam principle as this design largely compensates for the fluctuations in the source intensity. Sensitivity of the measurement will also depend on the proper selection of the filter.

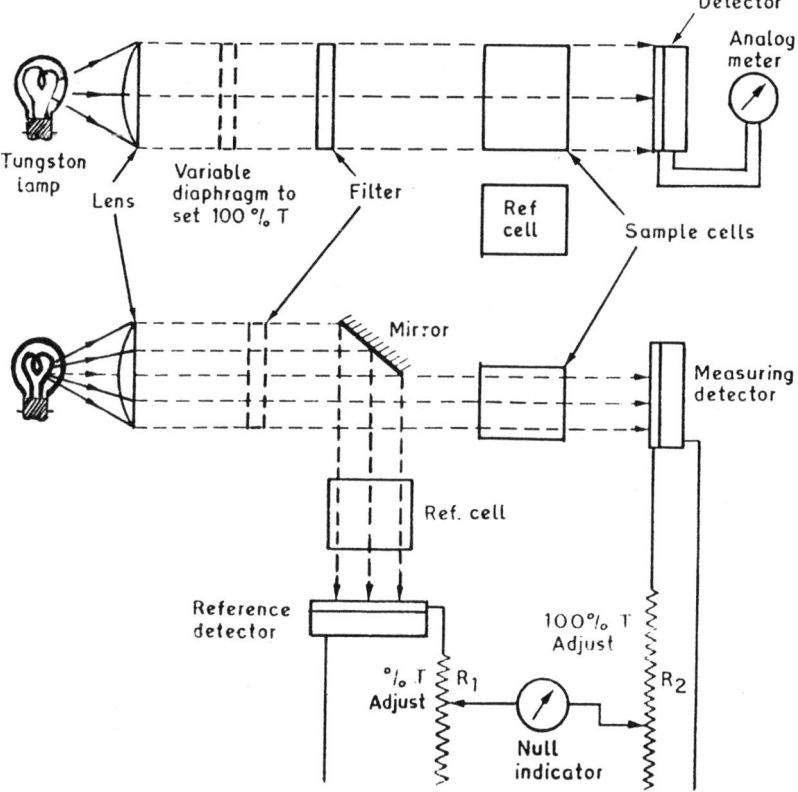

Fig.3.10 : Schematic diagrams of
(a) single beam, and
(b) double beam filter photometers

3.4.2 Single channel spectrophotometers

The inclusion of monochromator as a wavelength isolation device distinguishes a spectrophotometer from a filter photometer. A large number of single beam or single channel instruments are available from commercial sources which can be used for both ultraviolet and visible measurements. They are generally designed to operate in the general region 380 to 800 nm. However, lower and upper wavelength extremes upto 200 nm and 1000nm respectively are also available in some cases. They are equipped with interchangeable tungesten and hydrogen or deuterium lamps. In general, the gratings are employed for dispersion and photocells or photomultipliers are used for detection. Some of them are equipped with digital readout, but others employ analog meters. Since most of them are manually operated, the single channel instruments are not suited for any qualitative analysis that requires observing the absorption spectra over a wavelength region. Commonly they are used for quantitative analysis.

In order to perform absorption measurements on a sample with single channel instrument, at least, the following steps are required.

(i)   The monochromator wavelength is set at a desired value with the help of a wavelength selector knob.

(ii) With the shutter, in front of the detector, closed, the indicator is set to zero reading.

(iii) With the reference (e.g., solvent) or blank in the optical beam, the shuter is opened, and the indicator is set to 100% transmittance.

(iv) The sample is brought in the optical path and the percentage transmittance of the sample is read.

Therefore, if the absorption measurements are to be performed over a range of wavelengths, the utility of a single channel design is limited. Apart from being a time consuming procedure, the variables, that affect the measurement may change over the period of single measurement, and are not under full control. In such cases, double channel design appears more desirable. One channel permanently accomodates a reference or a blank and the other the sample.

3.4.3 Double channel spectrophotometers

How should the single beam design shown in Fig.3.2 be modified for double beam operation ? For an economic and efficient arrangement,

(i) a single beam should be employed except in the sample reference area, and

(ii) the beam should be split as it enters the sample-reference area and be recombined at the detector. The beam may be split either in space or in time using an appropriate beam splitter.

What mode of comparing the two beams yields best results? Should the device be optical null type or electronic null type? Each one of these has its own advantages and disadvantages. They will be clear from the discussion that follows.

Optical null type instruments

Some spectrophotometers employ an optical null procedure for comparing the intensities of the sample and reference beams. In this procedure, the power of reference beam is reduced or attenuated, to match that of the sample beam as shown in Fig.3.11. The attenuator is commonly a metallic comb, whose teeth are tapered so that a linear relationship exists between the lateral movement of the comb and the decrease in power of the beam. In the instrument shown in Fig.3.11, the radiation from the source is split into two beams (note the splitting in space).

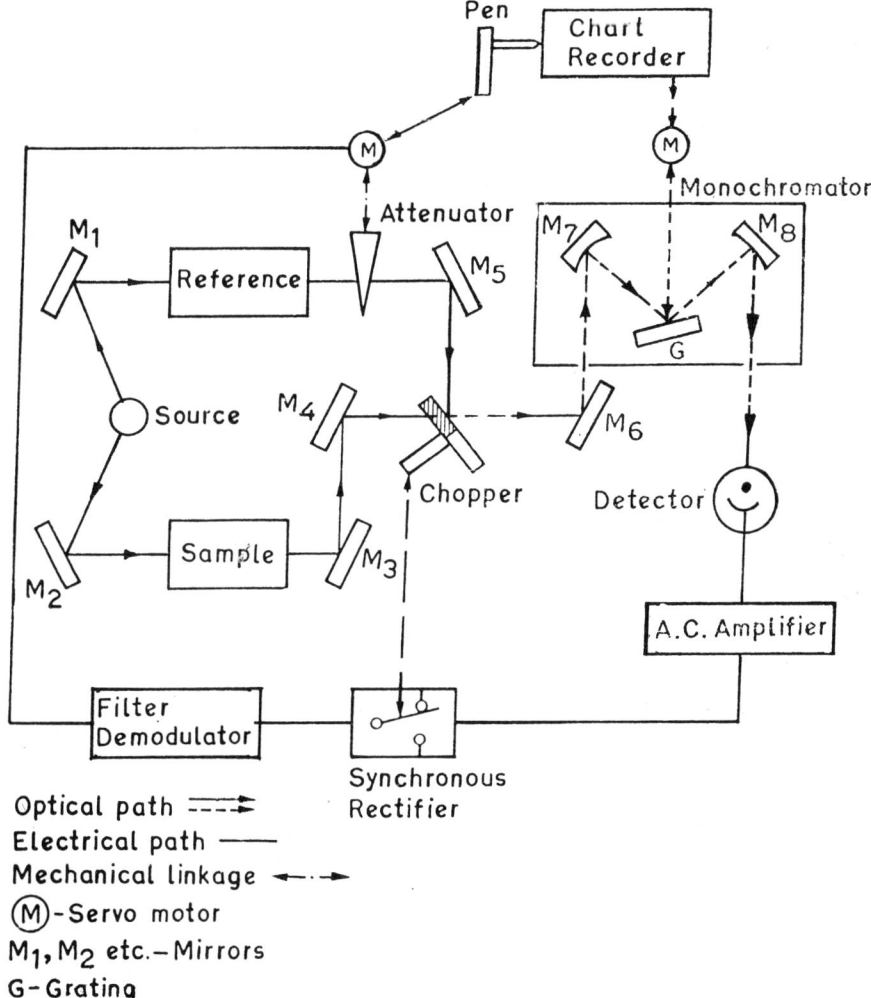

Fig.3.11: Schematic diagram of a double beam optical null spectrophotometer

Sample beam passes through an attenuator and onto a chopper. The chopper is a motor-driven rotating mirror sector that alternately reflects the reference beam or transmits the sample beam into the monochromator. After, dispersion, the alternating beams fall on the detector. The signal from the detector is amplified and passed to a synchronous rectifier which is mechanically or electrically coupled to the chopper. This arrangement causes both the rectifier switch and the beam leaving the chopper to change simultaneously. If the sample and reference beams are equal, the signal from the rectifier is d.c.;on the other hand, if the two beams differ in power, an a.c. signal is produced the polarity of which is determined by which beam is more intensive. The current from the rectifier is filtered and amplified to drive the synchronous motor in an appropriate direction. This motor is mechanically linked to both the attenuator and the pen drive of the chart recorder. The attenuator is driven up or down until the two beams become equal in power; or, in other words, an optical nulling is achieved. A second synchronous motor is used to drive the grating and the chart simultaneously.

In the optical null method, the detector is effectively utilized since it need determine only small differences in intensity. Secondly, a good S/N ratio is obtained and the components required are relatively inexpensive. However, there are some disadvantages which limit its use. The moving mechanical parts wear with time, inertia in movement causes slow response, and strongly absorbing samples are not accurately analysed as both the beams have low energy.

In general, the IR detectors have slow response times and hence most IR instruments employ optical null scheme with the beam being attenuated by a comb or an absorbing wedge. A photometric accuracy of $\pm 1\%$ is achieveable with a calibrated -wedge.

Electronic null type instruments

For obtaining greater precision in absorption measurements, particularly at high absorbances ($A_\lambda > 1$), an electrical or electronic nulling procedure has an edge over an optical nulling technique. The reasons are (i) there is inherent gain in accuracy with a long slide wire over the positioning of an atteunuator in a small cross-section beam, and (ii) even for high sample absorbance the energy in the reference channel is sufficient to operate the balancing motor efficiently.

Fig.3.12 shows, schematically, the design of double beam electronic null spectrophotometer. The optical beam from the source after passing through the monochromator is alternately directed via the sample and reference cells onto the detector.

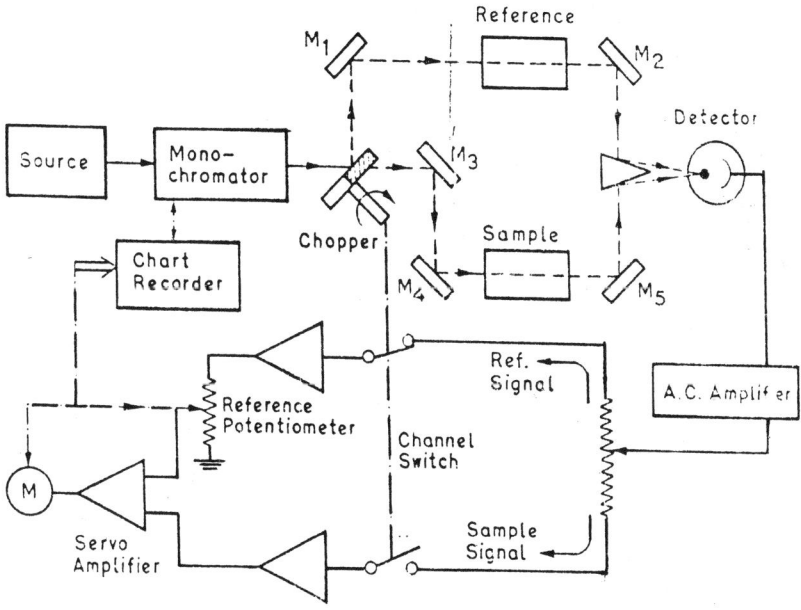

Fig.3.12: A schematic diagram of a double beam electronic null spectrophotometer

The beam-splitting is done with the help of a rotating sector mirror. The detector signal, after amplification, is directed by the synchronous channel switch to the sample and reference measurement circuits. The current in the reference channel generates a voltage drop proportional to the reference signal in the reference potentiometer. The signal from the sample channel is compared with the reference one. Any difference voltage is amplified by the servo-amplifier to provide a drive signal for the servomotor, M. The shaft of the motor is mechanically coupled to move the wiper of the reference potentiometer and also to move the pen of the chart recorder. The pen position on the chart is calibrated in terms of the reference voltage. There are other variations of this scheme, depending on whether the instrument is designed for manual or automatic operation.

Once the double channel is introduced into the instrument, it is worthwhile adding a few more servomotors and achieve automatic control as far as possible. In order to obtain an operational and calibrational stability for performing qualitative measurement the components that must be made automatic in operation are the wavelength drive, the slit control, the photometric balancing system and the readout device.

In the UV/VIS range, the radiation sources are, generally, quite energetic, the response of the detectors is much faster and narrow absorption bands are uncommon. Therefore, electronic nulling is much easier in UV/VIS spectrophotometers than in IR instruments.

3.5 APPLICATIONS

There are numerous inorganic and organic species that absorb in u.v., visible and i.r. regions; and are thus susceptible to quantitative as well as qualitative determinations.

However, u.v. and visible spectrometers have somewhat limited application for qualitative analysis because the number of absorption maxima or minima are very few in this region.

The trace analyses at 1 to 0.1 ppm level is possible with these instruments. Another application is in the structural investigations; that is, the information about the characteristic energy levels and the structural parameters of molecules can be obtained. The investigation of intermolecular interaction is also possible.

Depending on the instrument design, a single channel instrument can yield an accuracy of ± 0.5 to ± 2%. With double channel optical null spectrophotometers, the accuracy may be ± 1% if the optical attenuator has linear transmittance along its length and with electronic null instrument, the accuracy may be enhanced to ± 0.5%. Further improvement is possible with calibration or potentiometer compensation techniques.

TUTORIAL-3

3.1 Fig.3.13 shows a schematic diagram of a typical single channel reflection spectrometer (an instrument for recording absorbance through reflection technique) for measuring absorbances of a process fluid in the IR range. If the process fluid is a red hot metal, its blackbody radiation in the IR range will add to the measuring signal.

   (a) Suggest the modification in the design of this instrument to over come this problem.

   (b) How can the double beam operation be achieved? Should it be optical null or electronic null type device? Sketch the design.

   Hint: It should be noted that the radiation from the process fluid is continuos. Thus, in order to isolate the measuring signal from the background, the imposed signal from the source should be modulated. The detector will detect both the continuous background and the attenuated signal. By putting an ac amplifier of signal frequency, the background can be eliminated. The signal after amplification can be demodulated and readout.

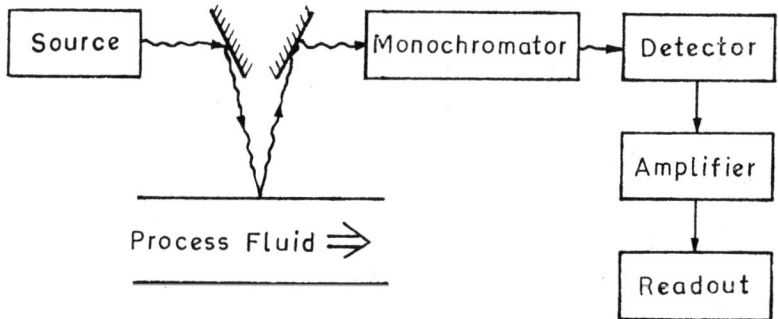

Fig. 3.13

3.2 An UV/VIS spectrophotometer is employing a tungsten filament lamp as a source of radiation that is coupled to the monochromator. The lamp is operated at 6000K. Assuming that the mechanical slit width of 0.05 mm gives adquate energy at λ_m and that the instrument is to scan from 0.4 to 1.0 μm.

(a) If the source intensity is taken as unity at λ_m, what is its relative value at 0.6 and 1.0 μm ?

(b) What should be the slit width at the above mentioned wavelengths in order to give a constant energy input to the monochromator ?

(c) The energy input to the monochromator can be varied either by varying the width of the entrance slit or by varying the current to the source. Which scheme should be preferred and why?

Answers : (a) 0.90, 0.374 (b) 5.56 μm, 13.3 μm

3.3 (a) In an interference filter, the dielectric spacer has a refractive index of 1.6. What should be the minimum thickness of the spacer, if it is to transmit the i.r. wavelengths ranging from 0.8 to 40 μm, in the first order?

(b) A Fabry-Perot interference filter that passes second order band centred at 0.30 μm is available. In order to eliminate other bands passed by the filter, an auxiliary filter is required. What wavelength should be selected for the first order band of the auxiliary filter for optimum blocking of bands (other than 0.30 μm) passed by the basic filter ?

Answers : (a)12.5 μm (b) 0.90 μm

3.4 An absorption spectrophotometer, covering the spectral range u.v.-visible and ir, is to be designed.

(a) What type of monochromator should be used for achieving higher resolution and less stray radiation?

(b) What should be the site of beam chopper in the overall design? Consider that it could be placed before the entrance slit or after the exit slit of the monochromator and that the detector output will be selectively amplified at the modulation frequency.

(c) Should it employ optical null or electronic null method for recording the sample transmittance? Give reasons.

(d) Sketch schematically the design of the proposed instrument.

3.5 A tuned laser source is available. Its tuning capacity extends from near u.v.through the entire visible region.suggest a possible design of a UV/ VIS absorption spectrophotometer employing this source.

3.6 A typical absorption photometer is employing a low-pressure mercury discharge lamp that emits mainly a resonance radiation of 2537 A°wavelength. How can this instrument be used to detect mercury vapour without employing a monochromator?

Sketch a block diagram of the instrument and explain the function of each block.

3.7 In a specific experiment, it is required to obtain a precise record of an absorption peak in u.v., whose, natural bandwidth (B) is $0.20\,\mu m$. If the reciprocal linear dispersion of an UV spectrometer is $0.25\,\mu m/mm$, what should be the mechanical slit width of the instrument ?

Solution

Assume that the monochromator has a triangular slit function as shown in figure below. The figure represents the intensity pattern at the exit slit of the monochromatorr when monochromatic light of wavelength, λ_o, is incident. Under these conditions, if the ratio the spectral slit width, S, to the bandwidth of the peak at half-height (B) is about 0.1, the spectrometer will record the absorption peak with a precision of $\pm 0.5\%$. Therefore, for adequate resolution, S/B = 0.1. In the present problem, $B = 0.20\,\mu m$ and hence $S = 0.1 \times B = 0.1 \times 0.20 = 0.02\,\mu m$. We know that spectral slit width (S) = mechanical slit width (W) X the reciprocal linear dispersion (D^{-1})

Therefore, $W = \dfrac{S}{(D^{-1})} = \dfrac{0.02\,(\mu m)}{0.25\,(\mu m/mm)} = 0.08\,mm$

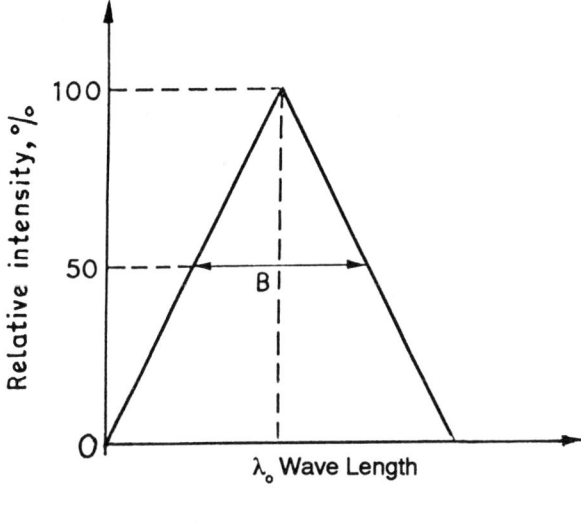

Fig. 3.14

Thus the slit opening of $W = 0.08$ mm may be used for adequate resolution.

3.8 An IR spectrophotometer has a spectral slit width of $1.25 \times 10^{-3}\mu m$ and a time constant of 0.1 second.

(a) What should be the maximum scanning speed so that the inherent resolution of the instrument is realized?

(b) What is the time required for scanning the absorption spectrum in the range 2.5 to 5 μm?

Solution

(a) The electronic components, e.g, detector, amplifier, or readout device require certain time interval to respond fully to the implied signal. The rate of response of the system is generally measured in terms of the time constant (τ) which is the time interval required for 63% or $(1 - 1/e)$ response when a sudden signal (step function) is received at the detector. The system's response is also measured in terms of 'response time' which is the time interval equal to 4 τ; within the response time 98% response will be achieved.

According to this criteria, the maximum scanning speed that can allow the inherent resolution of the instrument to be utilized, will be given by the relation: scanning speed = $S/4\tau$, where S is the spectral slit width. In the present problem $\tau = 0.1$ sec. and

$$S = 1.25 \times 10^{-3} \, \mu m,$$

$$\text{Speed} = \frac{1.25 \times 10^{-3}}{4 \times 0.1} = 3.125 \times 10^{-3} \, \mu m/sec.$$

(b) The spectrum ranges from 2.5 to 5 μm and hence the total scanning length = 2.5 μm. Therefore the time required for scanning the absorption spectrum

$$= \frac{2.5}{3.125 \times 10^{-3}} = 800 \text{ sec.}$$

$$= 13.3 \text{ min.}$$

4

FLAME EMISSION AND ATOMIC ABSORPTION SPECTROMETERS

4.1 INTRODUCTION

Flame techniques are based on the emission, absorption or fluorescence of electromagnetic radiation by atoms in the vapour state. When an analytical sample is introduced into the flame, the latter atomizes the sample, i.e., it furnishes the atoms of most elements present in the sample. There are three ways in which the atomic vapour so produced may be utilized for analytical purposes. These are illustrated in Fig.4.1.

In the first technique, known as flame emission spectrometry (See Fig.4.1(a)), the analytical sample is introduced into the flame where it is atomized. A fraction of the atoms is excited. Subsequent de-excitation gives rise to emission lines (λ_{em}) corresponding to different elements present in the sample. Since the atoms of most metals are easily excited by flame, this technique is widely applicable to trace analysis of metallic elements.

In the second method, called atomic absorption spectrophotometry and shown in Fig.4.1 (b), a beam of radiation (λ_{inc}) is allowed to pass through the flame containing the atomic vapour. The atoms in the flame absorb characteristic wavelengths (λ_{abs}). Thus the measurement of the absorption spectrum gives the information about the elements present in the sample. This method is also used chiefly as a quantitative method, widely applicable to metals.

In the flame fluorescence spectrometry, illustrated in Fig.4.1(c), the atomic vapour is irradiated with a characteristic wavelength (λ_{exc}). This excites the fluorescence in the atoms. The emitted light (λ_{em}) is measured through a suitable equipment.

Although all the three techniques are useful for the trace analysis of metals and certain other elements, only the first two have been commercially exploited. Both these techniques are effective at nanogram and even at the picogram level. For elements that can be easily excited, emission method is best while for elements that are difficult to excite, absorption method is preferable.

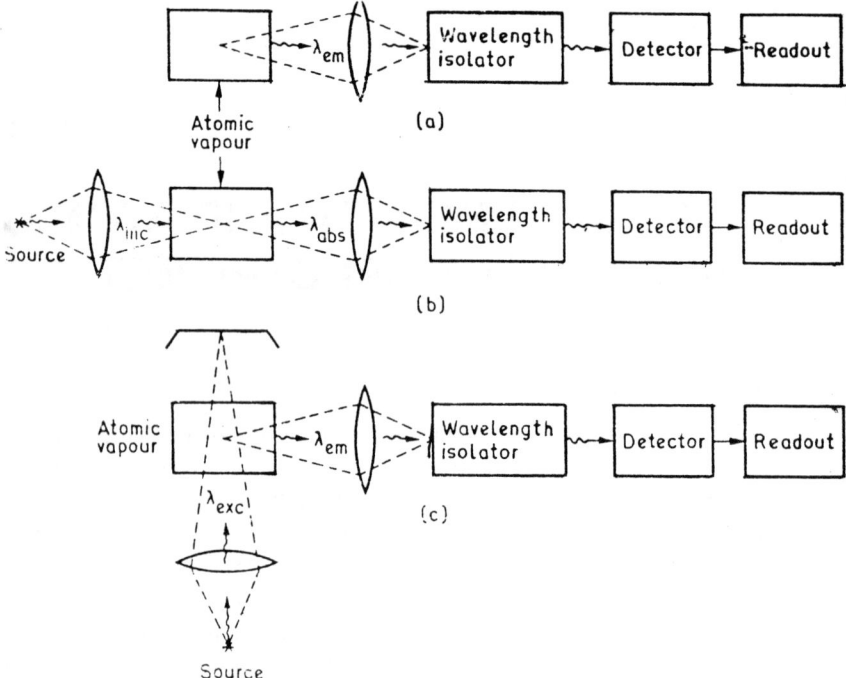

Fig.4.1 : Schematic diagram of an equipment for measuring
(a) atomic or flame emission,
(b) atomic absorption and
(c) flame fluorescence.

4.2 FLAME SPECTROMETRIC SYSTEMS: DESIGN CRITERIA

The instrumentation of all the flame spectrometric systems is nearly the same, as-illustrated in Fig.4.1. The techniques differ only in the modules used in the optical path prior to the flame. The flame serves as a sample cell as well as an atomizer and hence, it is often called a flame cell. In the following subsections we shall discuss the criteria for designing the systems based on the first two techniques, namely, flame emission and atomic absorption.

4.2.1 Flame emission spectrometers

A flame emission spectrometer (abbreviated as FES) can be represented block diagramatically as shown in Fig.4.2. On what factors does the signal developed by the detector depend? It has been shown* that the detector signal, S, may be given by the following expression.

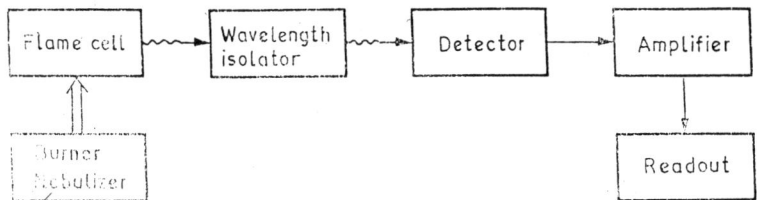

Fig.4.2 Schematic diagram of a FES

$$S = \frac{\phi \, \varepsilon \, \beta}{e_f Q} \, bhf_j \, A_{jo} \, C \left[\frac{N_j}{N_o} \right] \times M(\theta, \lambda) \times D(\lambda) \qquad (4.1)$$

Here N_j and N_o are the number of atoms of the analytical species in an excited and the ground state respectively; their ratio is given by

$$\frac{N_j}{N_o} = \frac{g_j}{g_o} \exp \left[- \frac{E_j}{k.T} \right] \qquad (4.2)$$

where k is the Boltzmann constant; T is the temperature in degrees Kelvin, E_j is the energy difference between the excited and the ground states; and g_j nd g_o are the statistical weight factors that are determined by the number of states having equal energy at each quantum level.

Other terms in eqn. (4.1) are identified as follows: ϕ is the rate of aspiration of the sample, ε is the efficiency of introduction of the sample into the flame; β represents the efficiency of atomization; b is the path length or thickness of the flame cell, Q is the flow rate of unburned gases, e_f represents a factor that takes care of the expansion of the sample volume in the flame; hf_j gives the energy of the photon corresponding to the transition from the excited j level to the ground level; A_{jo} is the Einstein coefficient for spontaneous emission, and C is the concentration of the analytical species in the solution. The factor M (θ, λ) represents the response of the monochromator in terms of the solid angle seen by it and its transmittance with wavelength and finally D (λ) gives the response of the detector as a function of wavelength. Main assumptions for the validity of eqn. (4.1) are that:

(i) there exists a thermal and chemical equilibrium in the central region of the outer cone of a premix flame,

*J.D. Winefordner and T.J.Vickers: "Calculation of the limit of detectability in atomic emission flame spectrometry", Anal Chem, 36,1939 (1964).

(ii) the flame is optically so thin that it does not self absorb or fluoresce, and

(iii) the radiation from sources other than the analytical species has been neglected.

What criteria can be formed about the instrument design from eqn. (4.1)? The linear dependence of S on the first three factors ϕ, ε & β alongwith the exponential dependence on temperature suggests that the design of a burner nebulizer system for obtaining a specific flame is a crucial instrumental decision. The background radiation may cause a serious error, and hence it must be taken care of preferebly through the instrument design. Finally, the choice of the monochromator and detector should be such that they respond efficiently and almost uniformly over the spectral range of interest.

4.2.2 Atomic absorption spectrometers

In the instrument shown in Fig.4.2, if we add a radiation source and an appropriate optics prior to the flame it becomes an atomic absorption spectrometer (AAS). How can its performance be related to the modular performance? The signal, S, developed by the detector can be given by the expression,

$$S = P_{o\lambda} \exp(-a'bc') \, M(\theta, \lambda) \times D(\lambda), \qquad (4.3)$$

where $P_{o\lambda}$ is the power furnished by the source. As the beam of radiation passes through the flame cell, it will be attenuated according to the Beer's law:

$$P_\lambda = P_{o\lambda} \exp(-a' \, bc'),$$

where P is power transmitted by the flame cell, b is the path length inside the flame and the terms a' and c' are absorptivity and molar concentration in the gaseous phase. The terms a' and c' are given by the following expressions:*

$$a' = 0.037 \, \lambda_o^2 \quad \frac{g_j}{g_o} \; . \; A_{jo} \; \frac{\delta}{\Delta \Gamma_D} \qquad (4.4)$$

$$\text{and} \quad c' = \frac{C \, \phi \, \varepsilon \, \beta}{e_f \, Q} \qquad (4.5)$$

---

* J.D. Winefordner and T.J. Vickers: "Calculation of the limit of detectability in atomic absorption flame spectrometry", Anal Chem,36,1947 (1964).

The new terms are identified as follows. λ_o is the wavelength at the centre of the absorption line, Δf_D, is the Doppler half width of the absorption line, δ is the factor that represents line broadening due to causes other than Doppler broadening and the numerical coefficient (0.037) includes a variety of factors. The set of terms equivalent to c' gives the concentration of atoms of analytical species in the flame. These terms also appeared in eqn (4.1). As usual $M(\theta, \lambda)$ and $D(\lambda)$ represent the efficiencies of the monochromator and the detector respectively. Thus the design considerations that apply to FES are also valid for AAS. However, a special challenge arises in AAS for tackling the problem of background radiation. This background radiation generally arises from several sources e.g., the luminosity of the flame, stray radiation and the emission by some atoms of the analytical species, that are excited by the flame or the incident radiation. The widely used solution to this problem in AAS is either to modulate the radiation source or to chop the beam radiation. Since the flame emitted light is steady, a selective amplification of the detector output can be employed to block the background radiation. Thus AAS will respond only to the transmitted light from the source. This approach is illustrated block diagramaticlly in Fig.4.3.

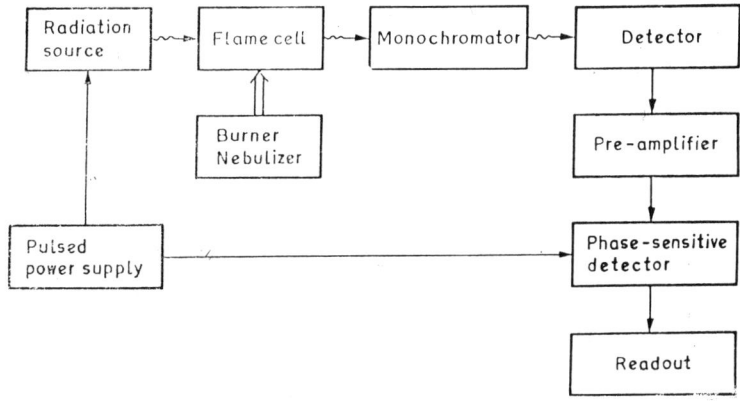

Fig.4.3 : Schematic diagram of single channel AAS

The signal from the atomic source, such as a hollow cathode lamp is either a.c. or a modulated d.c. The measurement information in the form of a signal attenuated by the flame cell is also accompanied by the steady background radiation. Both these i.e. the measuring signal as well as the background radiation are transmitted by the monochromator and detected by the detector. However, only the a.c. or the modulated d.c. signal is amplified and finally demodulated by the phase sensitive detector. The resulting d.c. is presented by the readout device.

In addition, if it is required that the signal be independent of the source variation, a double channel design must be employed. Further discussion in this regard may be seen in Sec.4.4.1.

4.3. COMPONENTS FOR FLAME INSTRUMENTS

As discussed earlier, except for the source, the components for both the flame instruments i.e. FES and AAS are common. Some of these e.g., monochromators, detectors, etc. have already been discussed in earlier chapters. Here we shall elaborate only those modules which are typical of flame instruments. Thus the burner-nebulizer system, hollow cathode sources, and resonance monochromators have been dealt with here.

4.3.1 Burner-nebulizer system

In most cases, the analytical samples are prepared in the form of solutions. The solution is aspirated through the capillary and a spray of droplets is produced by nebulizer. This spray is fed into the burner where during passage through the flame, the sample is evaporated and gets dissociated into free atoms. Thus a burner nebulizer system forms an important part of all the flame instruments. This point is also highlighted in Sec.4.2.1.

What should be the criteria for designing a burner nebulizer system? The system should be such that it satisfies the following requirements.

(i) It should produce a flame that is stable and has a well defined structure and shape. The luminosity, absorption and flicker should be least.

(ii) It should be able to nebulize as large a fraction of the sample solution as possible and at the same time eliminate large droplets of solution before the spray reaches the flame.

(iii) For absorption analysis, it should produce a flame that provides a large population of atoms. For emission analysis, the flame should also be able to excite the atoms of the analytical species.

(iv) The mechanical design should be such that it ensures rapid response to new samples, cleaning and adjustments are easy and above all prevents flashback into the burner.

(v) The speed of propagation of the flame front,generally called the burning velocity of the gas mixture, is also dependent on the burner design and other conditions such as pressure. Commonly, a steady flow of fuel and oxidant gases is achieved by employing a two stage regulator for each.

There are two types of burner nebulizer systems that are commonly employed in flame instruments. They are (i) total consumption burner and (ii) premix burner. These are shown in Fig.4.4 a and b respectively. In the first type, the sample is aspirated through a capillary directly into the flame. The fuel and oxidant gases mix by diffusion in the combustion zone and a small flame of circular cross section results. The advantages of this burner nebulizer system are that all the sample aspirated by the capillary appears in the flame, it responds quickly to new samples, is compact, mechanically simple and better suited to microsampling and offers no explosion hazards. The disadvantages are that (i) it provides a smaller flame that is unsuitable for absorption work, (ii) the flame has greater flicker, is luminous, turbulent and noisy; and (iii) the flame temperature changes drastically (from 100 to 400°C) when the sample droplets appear in it. This type of burner is mainly employed in emission work.

In the premix burners, the fuel and oxidant gases as well as the sample aerosol are blended prior to the emergence from the burner head. As the aerosol passes through the mixing chamber, the larger droplets are intercepted by the baffles and are drained away. This type of burner yields a laminar flame that has stable structure, low luminosity and little flicker. As the burner head has one or more flame slots that may be as long as 10 cm., the design is suitable for both emission as well as absorption analysis. The disadvantage is that there is a minor possibility of flashback explosion and hence the gas mixtures containing oxygen are not used with such burners. Instead, air or nitrous oxide is used as an oxidant with the fuel (e.g. acetylene). A carefully designed mixing chamber can reduce the volume of the mixing gases and lessen the danger.

4.3.2 Radiation sources: hollow cathode lamps

In the flame emission technique, the analytical species in the flame serves as a source of radiation; but for absorption measurements a separate source is always required. As the atomic absorption lines are remarkably narrow (0.002 to 0.005 nm) and as the electronic transition energies are unique for each element; the source requirements are highly specific.

The most common source employed in atomic absorption technique is the hollow cathode lamp. It consists of a tungsten anode and a cylindrical cathode sealed in a glass or plastic tube that is filled with an inert gas e.g. neon or argon (at a pressure of 1 to 5 torr). The cathode is made of a metal whose spectrum is desired or it serves to support a thin layer of that metal. The schematic configuration of this lamp is shown in Fig.4.5. When a potential difference of the order of 300 volts is applied across the electrodes, the inert gas atoms are ionized and establish a current of about 10 to 30 mA. The

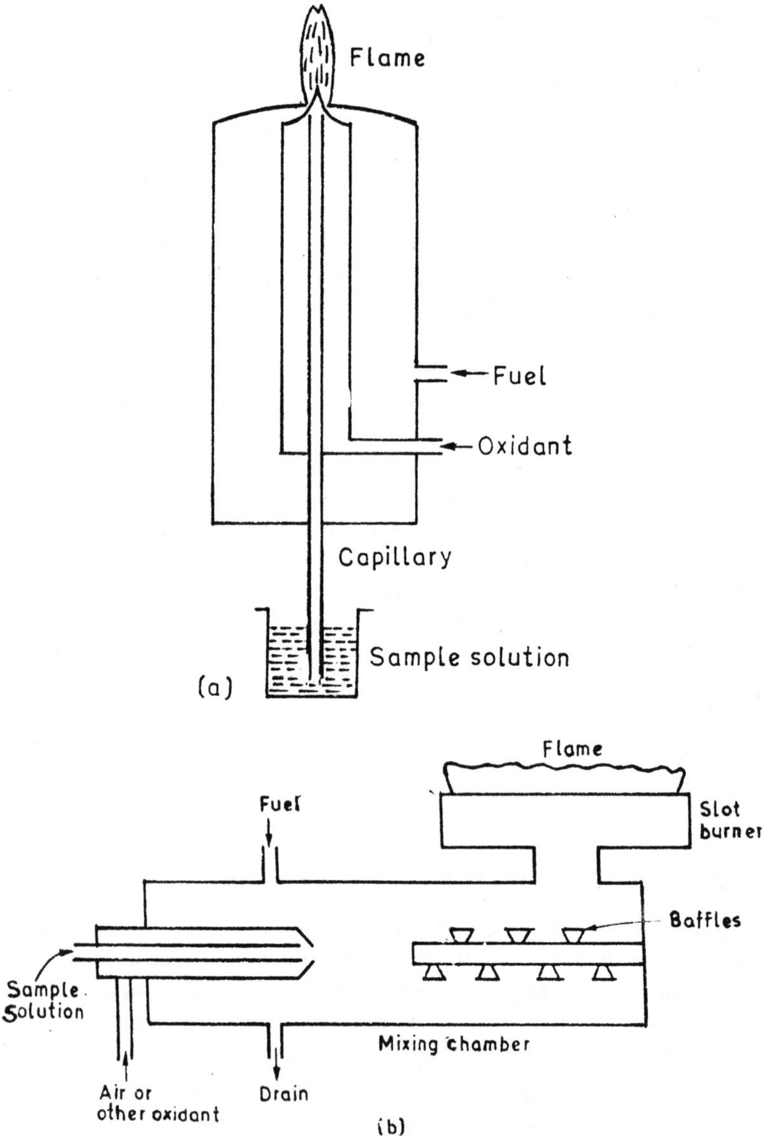

Fig.4.4 (a) Total consumption burner,
(b) Pre-mix burner

positive ions on their way to the cathode acquire sufficient kinetic energy and dislodge the metal atoms upon striking the cathode surface. Thus an atomic cloud of the cathode metal is produced. This process is called sputtering. A portion of the sputtered metal atoms get excited, and upon de-excitation emit their characteristic radiation. Eventually most of the metal atoms diffuse back to the cathode surface and are thus redeposited. The cylindrical configuration of the cathode helps in concentrating the emitted radiation in a limited region and also in enhancing the probability of redposition of the metal atoms on the cathode rather than on the lamp walls.

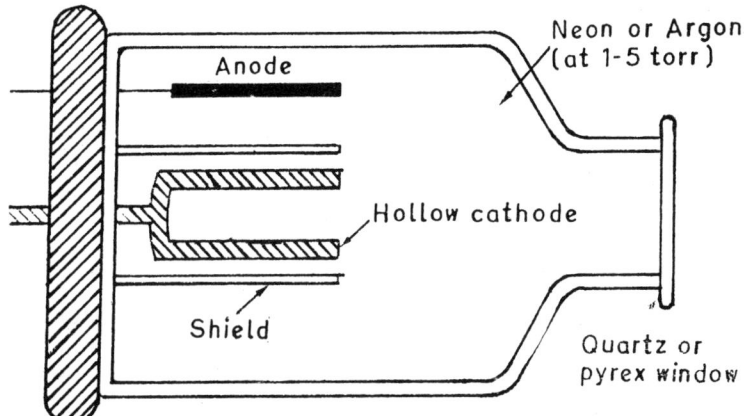

Fig.4.5 : Schematic diagram of a hollow cathode lamp

A variety of hollow cathode lamps is available, their efficiency depending largely on the geometry and the operating potential. In some cases, the cathodes are made of a mixture of several metals (5 to 6) which permit the analysis of all those metals employing a single source.

4.3.3 Wavelength isolation devices

Basically, it is required that the wavelength isolation device be able to separate the desired resonance line from the other lines emitted by a sharp line source, e.g. a hollow cathode lamp. For quantitative determination of alkli metals, glass and gelatin filters are sufficient, whereas for other metals with comparatively little spectra, interference filters are satisfactory. Most common and versatile wavelength isolation device is a conventional monochromator which can be set to pass any wavelength between 200 to 800 nm. Such monochromators have already been described in chapter 2.

An important class of wavelength isolators is encountered in atomic absorption measurements. It is called a resonance monochromator or resonance detector. It contains an atomic vapour of the element of interest, usually produced by cathodic sputtering (as in a hollow cathode source) or by direact heating. If the radiation from a line source such as a hollow cathode lamp containing the same element is allowed to pass through the atomic vapour, the resonance line (λ_{res}) or lines will be absorbed and reemitted as λ_{em} in all directions, whereas other wavelengths are not absorbed and pass through it unchanged. If this atomic vapour is viewed by a detector at right angle to the incident beam, the signal registered is found to be proportional to the intensity of the re-emitted resonance radiation, which in turn is proportional to the intensity of the resonance lines in the incident beam. An atomic absorption spectrometer, based on the use of a resonance monochromator is shown block diagrammatically in Fig.4.6. The effective slit width of such monochromator is on the order of 0.01 A°.If more than one element is to be detected, a dual atom resonance monochromator may be used. In this case, two filter-detector combinations will be needed at right angles to the incident beam; one for each element.

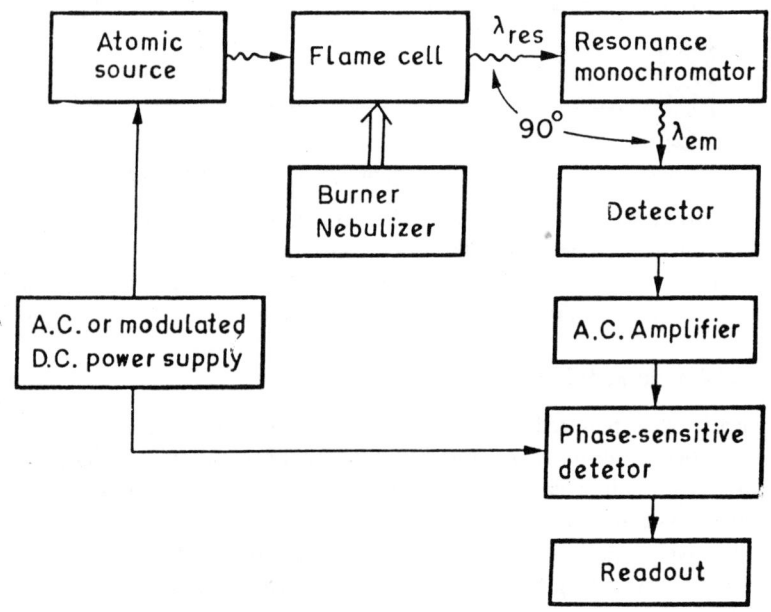

Fig. 4.6

4.4 FLAME SPECTROMETRIC SYSTEMS: SOME DESIGNS

4.4.1 Atomic absorption spectrometers

The instrument designed for atomic absorption measurements must be capable of providing a sufficiently narrow spectral band width to isolate the desired line from other spectral lines. For metals, which have a few widely spaced resonance lines in the visible region, a narrow bandpass filter is sufficient. However, most of the instruments employ a good quality uv-visible monochromator for spectral isolation. The detector and readout devices are similar to those already described for molecular spectroscopy in the uv/visible region. Both single and double beam designs are available.

As shown in Fig.4.3, a single beam instrument typically consists of a set of hollow cathode sources, a chopper or a pulsed power supply, a burner-nebulizer system and a simple grating monochromator with a PMT as a detector-transducer. The dark current is nulled with the shutter in front of the PMT. The blank or reference is first aspirated into the flame, and the 100% T adjustment is made. Finally the percentage transmittance is obtained with the sample replacing the blank. Fig.4.7 shows a schematic diagram of a typical double beam instrument. The optical beam from a hollow cathode lamp is split by a rotating sector mirror (chopper). Alternately the beam passes through the flame and the blank. The two beams are recombined at a recombiner and then passed on to the monochromator. Normally PMT serves as a detector, the output of which is fed into the lock-in amplifier that is synchronized with the chopper drive. The ratio of the sample and reference signals is amplified and fed to the readout device which may be a digital meter or a recorder. Alternatively, it is possible to attenuate the reference signal by means of a potentiometer to match the sample signal, thus measuring the transmittance or absorbance in terms of the position of the slide wire contanct.

4.4.2 Flame emission spectrometers

These spectrometers have found wide application in elemental analysis, particularly the determination of sodium, potassium, lithium and calcium in biological samples. The instruments for flame emission analysis are almost similar in design to the atomic absorpion spectrometer except that the flame now acts as a source of radiation; and hence, the hollow cathode lamp and the chopper are unnecessary. Most of the modern instruments are adoptable to both the emission and absorption analysis.

For a qualitative work or non- routine analysis, a recording uv/visible spectrophnotometer with a resolution of about 0.5 A° is desirable. For routine work or quantitative estimates, simple filter photometers are sufficient.

Some manufacturers supply flame photometers designed specifically for

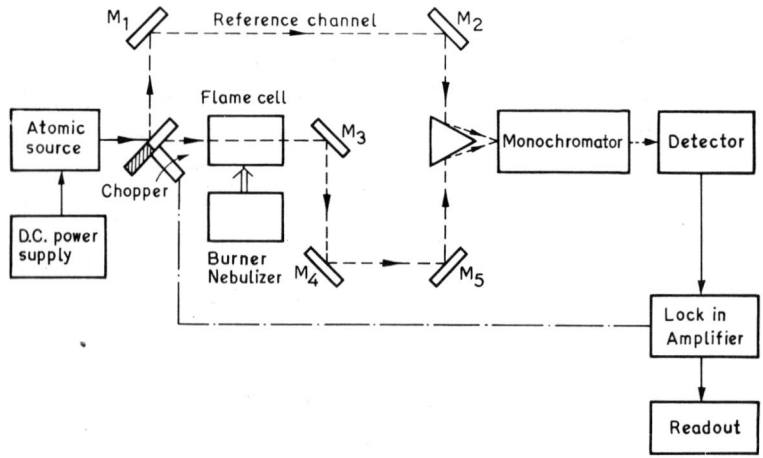

Fig.4.7: Schematic diagram of a double beam atomic absorption
spectrophotometer

the analysis of sodium, potassium and lithium in the biological samples. A
schematic diagram of one such instrument is shown in Fig.4.8.

In this instrument, the radiation from the flame is split into three beams
of approximately equal power. Each beam falls on a separate channel
consisting of an interference filter (which transmits an emission line of one of
the elements while absorbing that of the other two), a phototube and an
amplifier. Two channels are employed for measurement and the central
channel is used to monitor the internal standard such as lithium. The system
compares the output of each sample channel with that of the internal standard
channel through an electrical nulling procedure. The sample and reference
signals are summed, the sum amplified and is used to drive the servomotor in
a proper direction so that resultant sum becomes zero. The final readout is the
ratio of outputs of the sample and reference channels (corrected for dark
current). Such a system provides improved accuracy because the intensity of
each emission line is affected in the same way by the variables such as flame
temperature, flow rate of the fuel and the background emission.

4.5 ANALYTICAL APPLICATIONS

Atomic absorption spectrometers are sensitive instruments for the quantitative
determination of more than 60 metals or metalloid elements. The resonance
lines for nonmetallic elements lie below 200 nm and therefore, require
vacuum type spectrophotometers. The detection limits for many elements lie

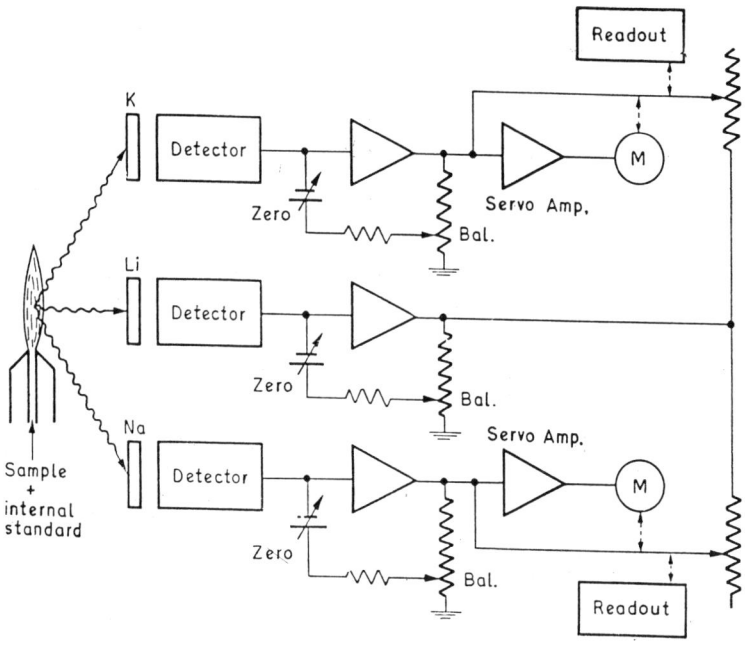

Fig.4.8 : Schematic diagram of three channel filter photometer

in the range of 1 to 20 ng/ml or 0.001 to 0.020 ppm.

Flame emission spectrometers find a widespread application in elemental analysis. This technique has been applied, of course, with varying degrees of success to the determination of about half the elements in the periodic table. Both the qualitataive as well as quantitative analyses are possible.

TUTORIAL-4

4.1 Referring to Fig.4.1(c), develop to criteria for designing a flame fluorescence spectrometer. On what factors does the signal developed by the detector, in this case, depend ?

4.2 Assume that the wavelength isolator used in the double beam atomic absorption spectrophotometer (shown in Fig.4.7) is a Czerny-Turner grating monochromator.

(a) In order to increase the wavelength scanning range of the instrument which of the following arrangements should be made ? Give reasons for your answer.

(i) A prism may be placed in series with a grating.

    (ii) Two gratings in series may be employed.

    (iii) Two gratings spanning two adjacent spectral regions may be mounted back to back.

  (b) Is it necessary to employ an order-sorting filter? If yes, what should be the site of the filter?

  (c) If the chopper is removed and the current in the atomic source is modulated, what change in the design of the instrument would be needed?

4.3 For a double beam AAS, suggest the remedies for the following problems that may be encountered. (You may think along the following lines,

    (i) the differential character of the double beam operation may take care of the malfunctioning.

    (ii) A calibration may be required,

    (iii) A change in the normal settings of the instrumment, e.g. sensitivity, gain, concentration etc. may be required

    (iv) The replacement of some component is necessary etc).

      (a) The backbround is present in the source radiation itself. Consider two possibilities; viz,

        (i) the background radiation is due to a resonance line emitted by some elemental impuritry present in the filler gas and this resonance line is near the resonance line emitted by the cathode, and

        (ii) there is a general overlap over the resonance line emitted by the source.

      (b) The sensitivity is poor.

      (c) The resolution is poor.

      (d) Optical alignment is not proper.

      (e) Some of the mirrors are fogged.

4.4 Discuss the merit of each of the following arrangements in an AAS, which is based on the use of resonance monochromator. Two elements are to be detected simultaneously. In each case, sketch the block diagram of AAS and show the optical path.

  (a) Single atom resonance monochromator is used.

  (b) Dual atom resonance monochromator is used.

4.5 How can a simple yet sensitive leak detector for mercury vapour be designed employing a resonance radiation of mercury at $\lambda = 2537A°$? What assumptions must be made in using this instrument?

5

FLUORESCENCE
SPECTROMETERS

5.1 INTRODUCTION

There is a special class of phenomenon, called luminescence, that is exhibited by a large number of materials in solid as well as liquid form. Luminescence has been defined as the emission of radiation by a sample of matter upon absorption of energy in some form. If the absorbed energy is in the form of photons of uv/visible light, the phenomenon in called photoluminescence. Similary, the emission caused by electrical excitation is called electroluminescence, that caused by high energy radiation such as X-rays or γ-rays is called radioluminescence, the one caused by chemical reaction is termed chemiluminescence, and so on. The processes of absorption of energy and its remission as light in all these cases are quite complex and their explanation is beyond the scope of this book. However, we discuss, in general terms, the relevent aspects of photoluminescence in the following paragraph.

Photoluminescence has been subclassified in two categories, viz.(i) fluorescence, which is the immediate release of energy as light (say, within about 10^{-8} sec or so) upon absortpion of light and (ii) phosphorescence, which is the delayed emission, i.e. emission after cessation of excitation. The time delay of phosphorescence may vary from about 10^{-3} sec to several seconds and even hours in some cases. A generalized energy level diagram depicting the possible transitions giving rise to these two phenomena is shown in Fig.5.1.

What are the steps involved in the emission of molecular luminescence? A molecule in the ground state (S_o) has its electron spins paired and the net spin is zero. The multiplicity of such a state is 'one' and hence it is called a singlet state. The absorption of energy by the molecule may raise it to the first excited state, S_1 without change in the spin. Such an excited state would also remain a singlet. However, if the electron spins become unpaired for the

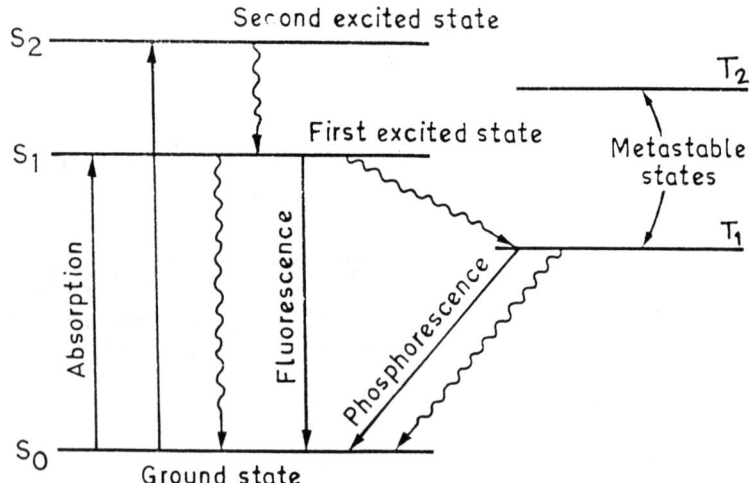

Fig.5.1: Generalized energy changes giving rise to fluorescence and phosphorescence. Wavy lines show non-radiative return transitions.

excited state, the multiplicity of the state becomes 3 and hence it is called a triplet, say, T_1. (This nomenclature is not followed in solid state luminescence exhibited by various impurity activated solids. In this case, the states S_o, S_1 S_2 etc are the energy levels of the activator centre and T_1, T_2 etc are known as the trapping levels). There are number of vibrational levels (not shown in Fig.5.1) associated with each of the states S_0, S_1, S_2 or T_1 and T_2. There are four ways through which an excited system may loose its excess energy. These are (i) the excitation energy may be lost in collision and in the process the system may come to the ground level through a non-radiative transition, (ii) the excited system returns almost immediately to the ground state with the emission of a photon (ie, emitting fluorescence), (iii) there is internal conversion of the lowest excited state (say, S_1) to a metastable state T_1 with a partial loss of energy (this process, in molecular luminescence terminology, being known as inter system crossing), followed by a radiative return transition from T_1 to S_o giving phosphorescence and finally (iv) the ransition from T_1 to S_o may again be non-radiative.

The fluorescence emitted by different analytical species, e.g. molecules or ions is characteristic of the species and hence this method forms an excellent tool for the identification as well as quantitative determinations of various substances in very dilute solutions. The method is effective even at ng/ml.

5.2 DESIGN CRITERIA

Before we discuss the design criteria of an equipment for measuring fluorescence, let us have a look at the power relationships involved, in their simplest form. Assume that the power incident on the sample solution at a particular wavelength λ is $P_{o\lambda}$. Its absorption by the sample will be governed by Beer's law. The transmitted power will be given by the relation $P_{\lambda} = P_{o\lambda} \exp(-abc)$, where a is the molar absorptivity at that wavelength, b is the optical path length in the sample and c is the molar concentration of species. The power that is absorbed by the sample may be expressed by the relation,

$$P_{o\lambda} - P_{\lambda} = P_{o\lambda}[1 - \exp(-abc)] \tag{5.1}$$

The intensity of fluorescence is directly proportional to the amount of radiation absorbed by the sample. As the quantum efficiency of fluorescence, ϕ_f, is defined as the ratio of the number of emitted photons to the number of absorbed photons, the power radiated by the sample as fluorescence, P_f may be given by the following relation.

$$P_f = \phi_f P_{o\lambda} [1 - e^{-abc}] \tag{5.2}$$

Of course, this power is radiated at a different wavelength. Normally, the wavelength of emission is shifted towards longer wavelength side as compared to the absorbed one. This is called stokes'law. However, there are cases, where the fluorescent emission is observed at shorter wavelengths. This is known as antistokes' shift.

With this understanding, what criteria should be formed for arriving at the design of the equipment for measuring fluorescence ? First, a polychromatic, preferably continuous, radiation source followed by a wavelength isolation device is needed for selecting the excitation wavelengths. A second wavelength isolator is required for analysing the fluorescence spectrum. This is to be followed by the detector-amplifier-readout assembly, as usual. A special challenge arises while designing the photometric system. The fluorescent emission is to be detected while the sample is being illuminated, i.e., excited. Thus a linear alignment of source-sample detector would cause severe problems as the radiation transmitted by the sample will be mixed with the fluorescent radiation. What solution exists for this problem? Use can be made of the fact that the fluorescence is emitted in all possible directions and hence, for the samples in the solution form, the fluorescent emission can be tapped at right angles to the beam of exciting radiation. (For another viewing mode, see tutorial 5.1). Accordingly, the modules in a fluorescence spectrometer may be arranged as shown block diagrammatically in Fig.5.2.

On what factors does the signal (S) developed by the detector depend?

From eqn(5.2) and considering the configuration of Fig.5.2, it follows that

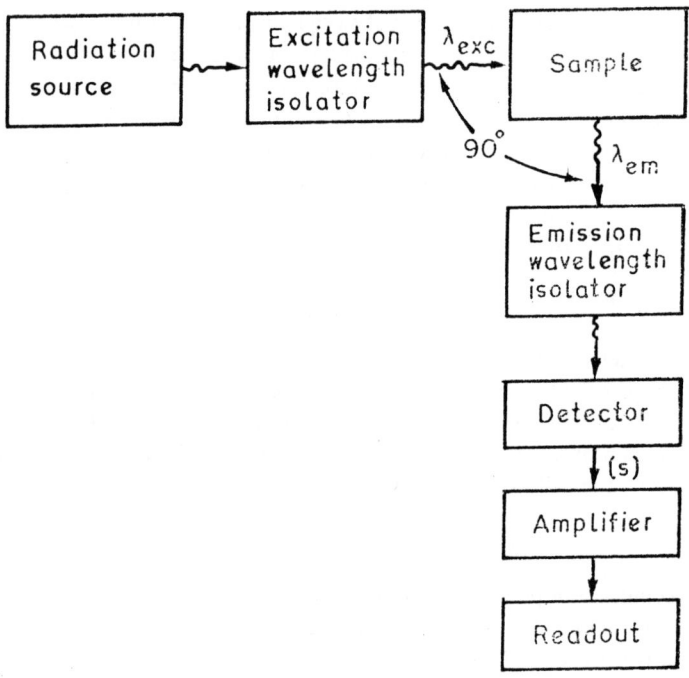

Fig.5-2 : Schematic diagram of a generalized fluorescence spectrometer (λ_{exc} is the excitation wavelength and λ_{em} is the wavelength of fluorescent emission).

$$S = \phi_f \, P_{o\lambda} \, (1 - e^{-abc}) \, M(\theta, \lambda) \times D\,(\lambda) \qquad (5.3)$$

where $M\,(\theta, \lambda)$ is a factor that governs the solid angle (θ) seen by the emission wavelength isolator and its transmission efficiency with wavelength, λ; and $D(\lambda)$ gives the response of the detector with wavelength, λ.

A linear dependence of S on $P_{o\lambda}$, the power of the exciting wavelength, requires that the excitation source should be quite intense. In order to have the choice of many wavelengths, the source should be polychromatic. Preferably, it should be continuous. The sample holder should be such that it transmits both the wavelengths, λ_{exc} and λ_{em}. With intense source, optical filters are excellent choice for wavelength isolation in quantitative work. In fact, a wide range of filter fluorimeters have come into existence. For qualitative investigations though, a good quality monochromator is required for isolating excitation as well as emission wavelengths. Such instruments are called spectrofluorimeters.

The optical arrangement should be such that the output beam from the excitation monochromator uniformly illuminates the sample and that the emission monochromator collects as much fluorescent radiation as possible. The response of the detector should be high and almost uniform over the desired wavelength range. These criteria should be kept in mind while selecting the modules for a fluorescence spectrometer.

5.3 THE CHOICE OF COMPONENTS

Source: Theoretically, any continuous source may be used for excitation purposes. In this connection, a tungsten filament lamp or tungsten-halogen lamp with a quartz envelope may seem ideal. However, such sources tend to be weaker at shorter wavelength side as compared to high pressure sources. Incandescent sources are, therefore, not employed. Instead, compact source arcs, discussed in sec 3.3.1 are quite common.

Mercury arc lamps are often used in filter fluorimeters and xenon arc lamps are employed in spectrofluorimeters. There are two reasons for this, viz. (i) the spectrum emitted by the xenon arc is continuous and (ii) the intensity available in the entire range is also high. The spectra radiated by typical quartz tungsten halogen (QTH) lamp , xenon(Xe) arc and mercury (Hg) arc lamps are shown in Fig .5.3.

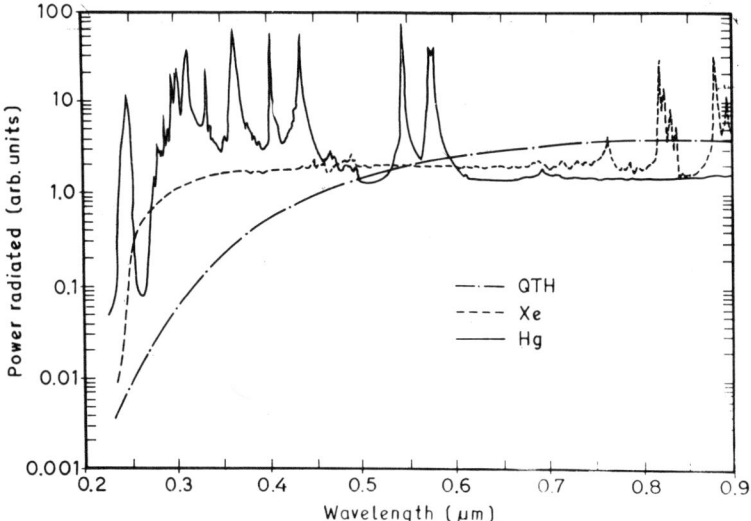

Fig.5.3 : The spectral distribution of power radiated by typical QTH, Xe and Hg lamps.

Wavelength isolator: In filter fluorimeters, with intense source such as a Hg-lamp, optical filters are sufficient for isolating the excitation as well as emission wavelengths. Color glass filters, discussed in sec.3.3.2, are normally used; though narrow band pass interference filters can also be used. For any qualitative investigation, however, a monochromator becomes a must. Spectrofluorimeters are, therefore, equipped with high resolution monochromators. Use of grating monochromators is common. Some instruments employ double monochromator for excitation as well as emission.

Detector: So long as the response of the detector in the measurement range is good, any detector may be used for measuring fluorescence intensities. However, most commercial designs are based on the use of photomultipliers for detection purposes.

Sample holders: For liquid samples, the sample holders, called cuvettes, are normally made of glass or quartz. They are either rectangular or cylindrical in shape. The design may vary for solid samples, depending on whether it is in the powder form or a single crystal form.

5.4 FILTER FLUORIMETERS

When the optical filters are used as excitation and emission wavelength isolators of Fig.5.2, the equipment is called a single channel or single beam filter fluorimeter. In such instruments, a stable high pressure mercury arc lamp is used as a radiation source. The spectral lines emitted by this source are quite intense and narrow (as shown in Fig.5.3). The excitation and emission wavelengths are selected by employing either color glass filters or sometimes, interference filters. The detector is normally a photocell that is connected to an analolg meter. With the shutter, in front of the detector, closed, the dark current is offset and a zero on the scale is obtained. With the solvent in the curvette, the shutter is opened and the meter is read. This reading gives the amount of scattered light reaching the detector. The calibration for quantitative measurements can be done with the aid of sample of known concentrations and plotting the concentration versus fluorescence intensity curve. The concentration of unknown samples can be found by matching their fluorescence intensities with this curve. So long as the power supply is stabilized and there are a few samples to be analysed, this instrument is to be preferred because it is compact, cheaper and its maintenance minimal. However, if the number of samples to be analysed is large, or the power fluctuations are not under full control, the calibration may not remain valid for a longer time. In such a situation a double beam design appears desirable.

A schematic diagram of a typical commercial double beam fluorimeter is shown in Fig.5.4, Herein, a part of the beam passed by the excitation or primary filter is diverted onto a reference photocell through a reduction plate. The rest of the beam falls onto the sample cell. The fluorescence emitted by the sample is filtered by two secondary filters placed at right angles to the incident beam and detected by the measuring photocells. The outputs of the measuring and reference photocells are compared by using a current bridge.

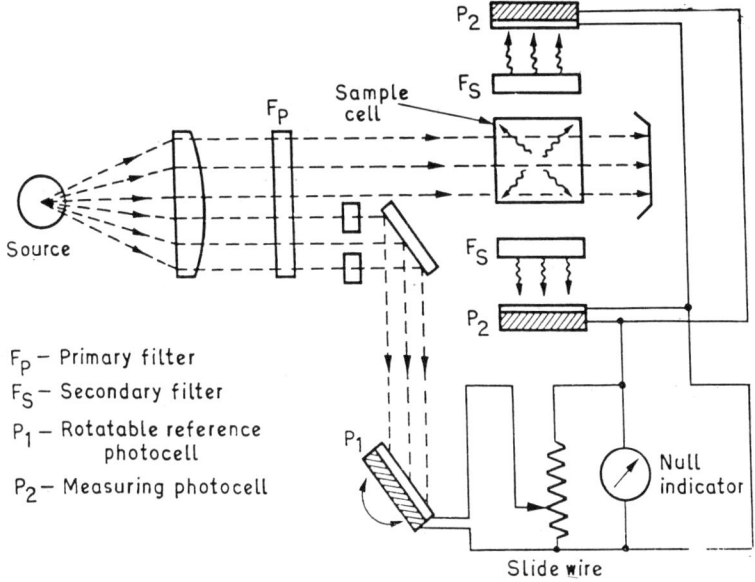

F_P — Primary filter

F_S — Secondary filter

P_1 — Rotatable reference photocell

P_2 — Measuring photocell

Fig.5.4 : A schematic diagram of a double-beam filter fluorimeter

5.5 SPECTRO FLUORIMETERS

If the qualitative investigations are to be performed or improvement in the accuracy of quantitative measurements is to be achieved, a good quality monochromator becomes a must both for excitation as well as emission. When the monochromatoors are employed in the design of Fig.5.2, it becomes a spectrofluorimeter or simply a fluorescence spectrometer. The instrument employing two Czerny Turner monochromators is shown schematically in Fig.5.5. Reflection gratings G_1 and G_2 used in these monochromators have typically 600 groves/mm with G_1 blazed for 300 nm and G_2 blazed for 500 nm in the first order. As usual, the filters are used for sorting out overlapping orders.

This type of instrument has the same drawbacks which appear with a single channel absorption spectrometer, as the sample readings are not compared with a standard or reference on a continuous basis. Thus, if the reference-sample compensation is required automatically and accurately, a double-beam design appears desirable. The optical beam from the excitation monochromator may be split by a rotating sector mirror, as shown in Fig.5.6.

The beam alternately strikes the sample and reference cells. The output of these cells is monitored at right angles to the incident beams. The optoelectrical arrangement for comparing the two outputs may be made as in the case of double beam absortpion spectrophotometers.

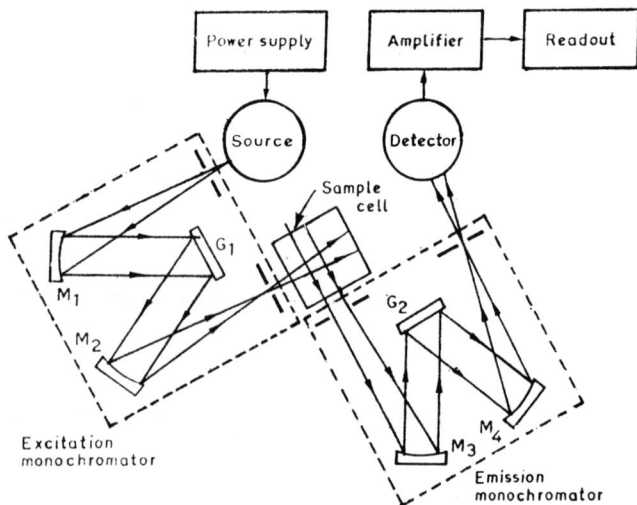

Fig.5.5 : Optical schematic of a single channel spectro- fluorimeter. G_1 and G_2 are gratings and M_1, M_2, M_3, M_4 are mirrors.

5.6 ANALYTICAL APPLICATIONS

The technique of fluorescence is quite sensitive as well as selective and hence it can be used for trace analysis of elements, ions and molecules. Over 50 elements of the periodic table can be easily analysed through this technique.

The molecular fluorescence is normally employed for quantitative analysis , though it can also be used for qualitative analysis.

Apart from use in analytical chemistry, the fluorescence spectrometers find wide application in many diverse disciplines; e.g., physics, forensic science, pharmacy, biology, medicine; geology and so on.

TUTORIAL : 5

5.1 If the analytical sample is a semitransparent material or a solid powder or a highly absorbing solution, the fluorescence emitted by the sample can not be viewed at right angles to the excitation beam. What should be the viewing mode in such cases? For this mode, suggest the design of a single channel fluorescence spectrometer. What are the possible advantages and disadvantages of this configuration ?

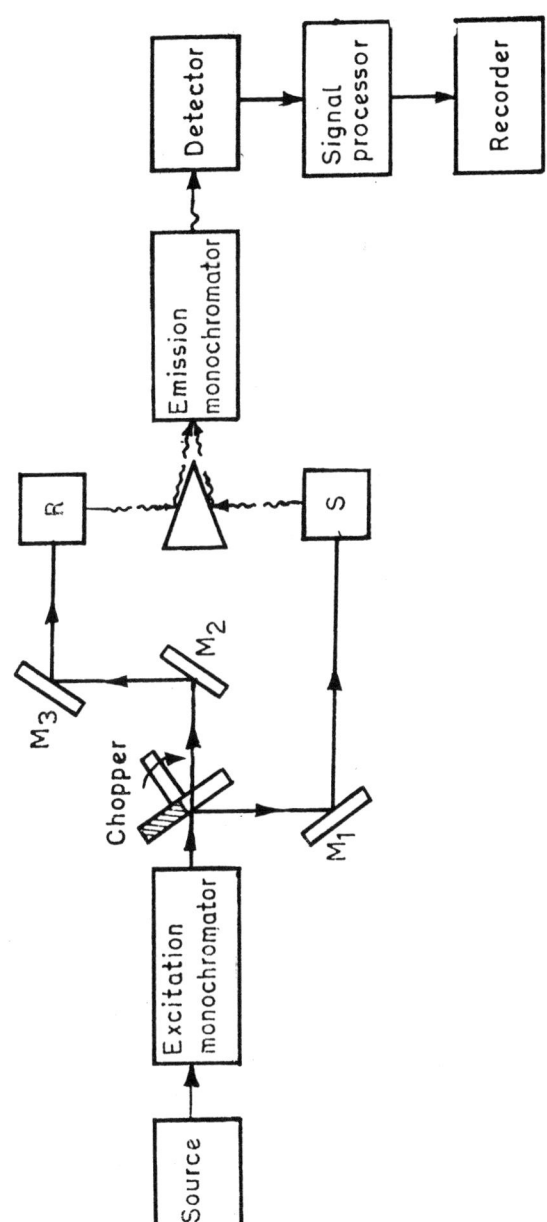

Fig.5.6 : A schematic diagram of a double-beam fluorescence spectrometer. R and S denote the reference and sample cells respectively.

Hint: The fluorescent emission may be collected on the same side on which the cell is illuminated. This is called frontal method of cell illumination. Approximately an angle of 37° between the excitation beam and the entrance axis of the emission monochromator has to be maintained for minimizing the reflected light.

5.2 Some analytical species (e.g., solutions, impurity activated solids, etc) emit light after the end of excitation. This is called phosphorescence (See sec 5.1). Suggest the design of the equipment for recording the phosphorescence spectrum.

Hint: Note that such a delayed emission will not be constant but will decay with time (according to power law or an exponential law). A xenon-flash lamp may be used for excitation. If the flash is made repetitive, a.c. methods of amplification may be used for detecting low intensity phosph orescence.

5.3 A high energy radiation, e.g., x-rays; may be used for exciting u.v./ visible fluorescence in certain inorganic solids.* Suggest the design of the instrument for measuring the fluorescence of such systems. (Note that the energy of x-ray photons is more than sufficient for exciting all the energy levels of the species under test and hence an excitation monochromator, in this case, is not required. This type of fluorescence is different from x-ray fluorescence emitted by elements. It is discussed in Chapter 6).

5.4. Suggest the steps involved in recording the excitation and emission spectra with the help of a spectro- fluorimeter. What measures, instrumental or otherwise, should be taken to obtain the true spectra?

---

*See for example, R.P.Khare and J.D.Ranade : 'Fluorescence of CaO phosphors activated by Ce and Tb under X-ray excitation; Indian J.Pure & Appl. Phys;13,664 (1975).

X-RAY ANALYSERS

6.1 INTRODUCTION

There are three different techniques of X-ray analysis. They are :

    (i) X-ray fluorescence

    (ii) X-ray diffraction, and

    (iii) X-ray absorption.

The fluorescence analysis enables to determine the presence of element(s), in a given sample from the wavelength and intensity of x-rays that are emitted from the sample after it is bombarded with other x-rays (of higher energy). As the x-rays emitted by the excited element have characteristic wavelengths of that element and the intensity of the spectral lines is proportional to the number of excited atoms, this method can be used for both qualitative as well as quantitative elemental analysis.

The diffraction analysis is based on the diffraction of x-rays from the planes of a crystal or a crystalline powder (to be analysed). This method enables to identify the crystal structure of the analytical species. Comparison of this structure with standards helps in identifying the species themselves.

The third method, i.e. absorption analysis, is based on the fact that different materials have differing absorption of x-rays and that sharpness in absorption occurs, for x-ray wavelengths, for which the energy of x-ray photon is nearly equal to that required for the removal of an electron from inner shells of an atom.

The instruments based on the three techniques are called x-ray fluorescence spectrometer, x-ray diffractometer and x-ray absorption spectrometer respectively.

6.2 THE BASIC PRINCIPLES OF X-RAY GENERATION

In a discharge tube shown in Fig.6.1, if the cathode is heated, it starts emitting electrons. These electrons get accelerated towards an anode if a sufficient potential difference exists between the two electrodes. Upon striking the metallic target (affixed to an anode) the electrons transfer their energy to its surface. The latter, in turn, gives of radiation which has very short wavelength $(0.1$ to $100\,A^0)$ and is called x-ray radiation. The wavelength of the generated x-rays depends on the target material and on the potential difference between anode and cathode.

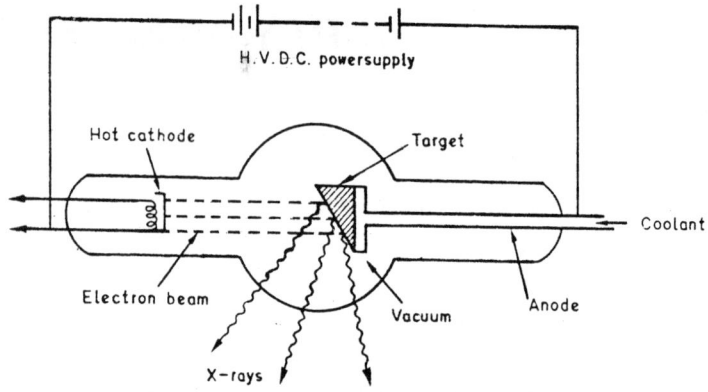

Fig.6.1: Schematic diagram of an X-ray source

What is the shortest wavelength that is attainable with a particular target? This calculation is rather straight forward. If it is possible to convert the entire energy of the electron impinging on the target into the radiant energy, we shall have x-rays of highest possible energy and hence of shortest possible wavelength.

The radiant energy of one x-ray photon is equal to hf, where h is the planck's constant and f is the spectral frequency of radiation; whereas the energy of one electron which has been accelerated through a potential difference of V volts, would be eV, (e being the electronic charge). If these two are equal, we have

$$f = f_{max} \text{ and } hf_{max} = eV$$

Since $\quad f = \dfrac{\text{Speed of light in vacuum}}{\text{The wavelength of the radiation}} = \dfrac{c}{\lambda}$

we have $\quad f_{max} = \dfrac{c}{\lambda_{min}}$

Thus, $\quad \dfrac{hc}{\lambda_{min}} = eV$

or $\lambda_{min} = \dfrac{hc}{eV}$

As $h = 6.6256 \times 10^{-34} Js$

$e = 1.602 \times 10^{-19} C$

and $c = 3 \times 10^8 \ m\text{-}s^{-1}$,

we have $\lambda_{min} = \left(\dfrac{12400}{V}\right) \times 10^{-10} \ (m) = \left(\dfrac{12400}{V}\right)$ (A°) (6.1)

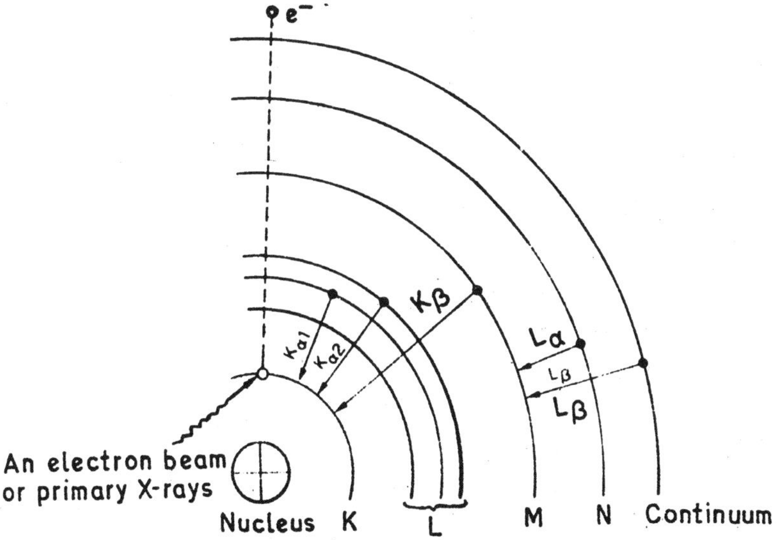

Fig.6.2 : Excitation of an atom by an electron beam or primary x-rays and some possible downward transitions.

This is the minimum attainable wavelength, λ_{min}, (also known as "short wavelength limit") of the x-rays that can be radiated by the target and is inversely proportional to the applied voltage (V). What is the mechanism responsible for the emission of x-ray spectra? We know that an atom is composed of a central nucleus and numerous electrons arranged in various shells (e.g. K,L,M, and so on); as shown in Fig.6.2.

When the atom bombarded by a highly energetic electron beam or energetic x-rays (which we can call, primary x-rays), the energy of the latter (i.e. the electron beam or primary x-rays) may be absorpbed by the atom. If this energy* is sufficient to knock out an electron from the inner shells (e.g. K or L shell), the atom becomes ionized. The vacancy of this electron is promptly filled up by an electron from a higher shell. This transition from higher to lower shell gives rise to emission of x-ray radiation. The energy of this radiation is obviously less than that of the agency which caused it.

Suppose an electron is displodged from a K-shell, it may be replaced by an electron from L or M shell. The transition from L-shell to K-shell will give rise to x-ray photon of energy,

$$hf = E_L - E_K \qquad (6.2)$$

where E_L and E_K are the energies of the electron in L and K shells respectively. The spectral lines that originate from such transitions are called K_α lines; whereas those originating from transition between M and K shells are called K_β - lines. Likewise, transitions from M or N shells to L shell are termed L_α or L_β lines respectively.

From this discussion, it would seem that the x-ray emission spectra is similar for all elements, because K_α, K_β etc lines are involved in all of them. This is not so, however. This nomenclature remains same but the actual wavelengths of these lines depend on the atomic number of the element. The relationship between the wavelength (λ) of x-rays and the atomic number (z) of the element is given by Moseley's law, which states that

$$\frac{c}{\lambda} = a (z - \sigma)^2 \qquad (6.3)$$

where c is the speed of light in vacuum, a is a constant and σ is also a constant dependent on the series of lines (e.g. K_α, K_β or L_α lines). A typical x-ray spectrum produced by the x-ray tube employing Mo target is shown in Fig.6.3.

6.3 X-RAY FLUORESCENCE SPECTROMETER

Suppose a sample (containing some elemental impurities) is placed in a beam of primary x-rays, they will be absorbed by the elements present in the sample. The absorbing atoms will be excited and their de-excitation will give rise to emission of x-rays with wavelengths characteristic of the elements. This process, of emission of secondary-x rays upon absorption of primary x-rays, is called x-ray-fluorescence. Since the wavelength of the fluorescent emission is characteristic of the element being excited, its measurement enables to identify the fluorescing element. The intensity of the spectral line is proportional to the concentration of the element in the sample and hence its measurement makes possible the quantitative determination of the element.

---

* The energy corresponding to the removal of an electron from K and L shell is called K_{edge} and L_{edge} respectively.

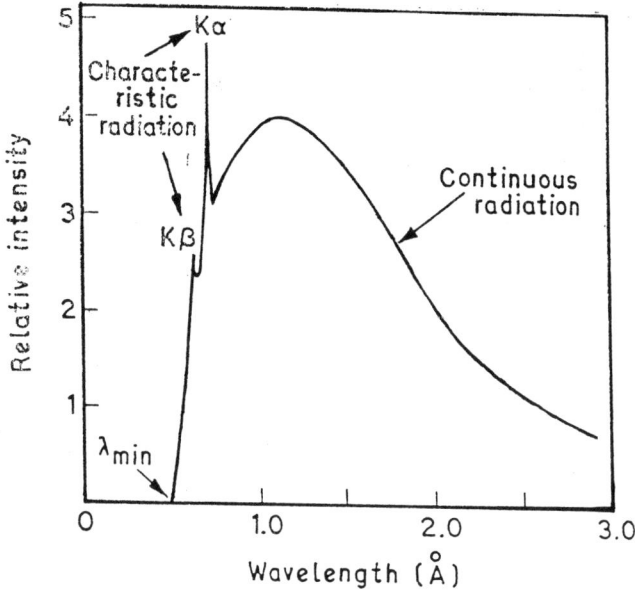

Fig.6.3 A typical x-ray spectrum of Mo

The instrument, that is based on this principle, is called x-ray fluorescence spectrometer.

6.3.1 Design criteria

What should be the criteria for designing the equipment for x-ray-fluorescence analysis?

For such an analysis, the first requirement would be a source of primary-x-rays. Secondly, the spectrum of x-rays emitted by the sample should be resolved by an appropriate monochromator and the corresponding wavelengths detected by the detector. The signal from the detector has to be processed as required and finally readout or recorded. Accordingly, the modules for x-ray fluorescence spectrometer may be arranged as shown in Fig.6.4.

The importance of each of these modules may be understood if one investigates the dependence of the detector signal(S) on pertinent variables.

$$S = [\, (1\text{-}R) \, QC \,] \, \phi \, (I_\lambda) \times M(\theta, \lambda) \times D(\lambda) \tag{6.4}$$

The terms in equations (6.4) are identified from left to right as follows: R is the fraction of the beam that is reflected by the sample; Q is the rate of excitation of the analytical species (e.g. the element present in the sample), C is the concentration of the analytical species; ϕ (I_λ) is a theoretical factor that determines the intensity (I_λ) of a particular spectral line of wavelength (λ); M(θ, λ) is the geometrical factor which determines the solid angle observed. by the monochromator and its transmittance with wavelength(λ) and, finally, D (λ) is the response of the detector with wavelength . Thus, while designing the x-ray fluorescence spectrometer R should be minimized and M(θ,λ) and D(λ) should be maximized.

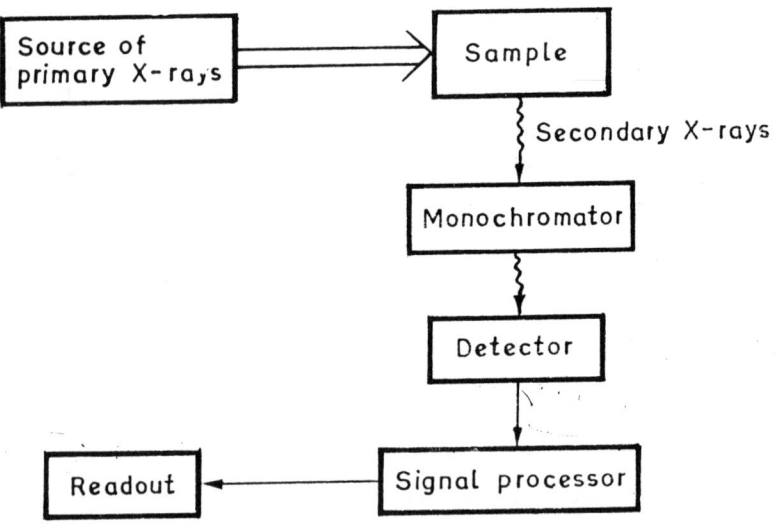

Fig.6.4: Block diagram of a generalized x-ray fluorescence spectrometer

6.3.2 Basic components

(a) X-ray source

A typical x-ray source has already been illustrated in Fig.6.1. The prime requisities of this module are :

    (i) it should provide x-rays of energy higher than that required for the excitation of the element being examined in the sample; and

    (ii) the source intensity and the spectrum of x-rays should be constant.

    The first requirement is generally met by the appropriate choice of the target element. As a rule of thumb, the atomic number of the target element should be higher than that of the element being examined. The second requirement is often met by supplying a highly stablized electrical power to the electrodes of the x-ray tube. During the operation of the source, the target gets excessively heated and hence some arrangement for its cooling is also required.

(b) Collimators

The x-rays emitted by the target are randomly oriented in all directions. As a result, the wavefront of x-rays forms a hemisphere with the target surface at the centre. In order to obtain a narrow beam of x-rays, two sets of lead plates are placed in the path of x-rays, as shown in Fig.6.5. These plates absorb all the radiation except that passing through the collimator gap. This arrangement serves as a collimator for x- rays.

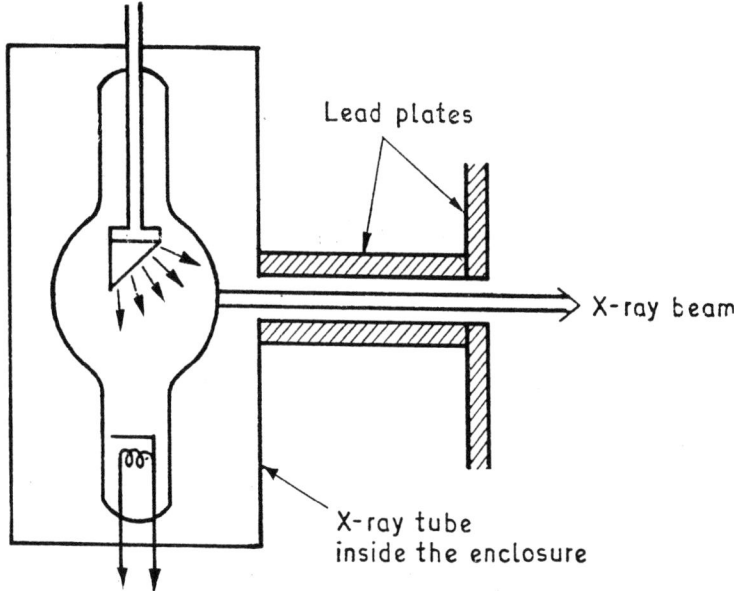

Fig.6.5: Collimating arrangement for x-rays

(c) Filters

A filter is a thin foil of a material that transmits x-rays selectively. For example, if the target in the x-ray tube is of copper (Cu), the x-ray emission spectrum would consists of Cu - K_α and K_β lines superposed on a continuous background depending on the voltage applied to the target. If the nickel (Ni) foil is placed at the window of the x-ray tube, it will absorb strongly the radiation at shorter wavelength but will absorb weakly the Cu K_α- line. (The relevent spectra is shown in Fig.6.6). Thus the radiation passing through the Ni filter comprises mainly of Cu- K_α - line and a minor background.

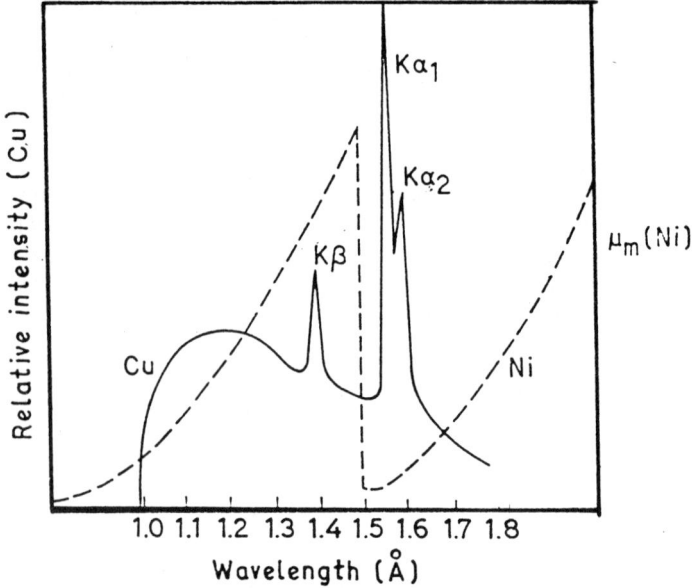

Fig.6.6:(a) X-ray emission spectrum of Cu-target (solidline) and
(b) Absorption curve of Ni filter (dashed line).

(d) Monochromator

In x-ray fluorescence analysis, a crystal of known lattice parameters, is employed as a dispersion device. The dispersion is achieved through the diffraction of x-rays by the crystal. Consider a set of two parallel rays X_1 D and X_2 B that are reflected (as DX_1' and BX_2') by two successive crystal planes (shown in Fig.6.7). The two rays travel the same distance upto AD and beyond CD (assuming AD and CD to be normal to the incident and reflected rays); but the second ray X_2BX_2' travels a distance AB + BC greater than the first one. Thus a path difference equal to AB + BC = 2 AB has been introduced between the two rays. Since AB = DB sin θ = d sin θ, where d is the interplanar spacing and θ is the glancing angle or the Bragg angle (the angle which the incident rays make with surface of the crystal), the path difference 2AB = 2d sin θ. If this path difference is an integral multiple of the wavelength (λ) of x-rays, they will re-inforce each other and produce a diffraction maxima.

Thus $\qquad$ $2d \sin \theta = m\lambda$ $\hfill$ (6.5)

where m is an integer (1,2,3, etc) and is known as an order number. Eqn

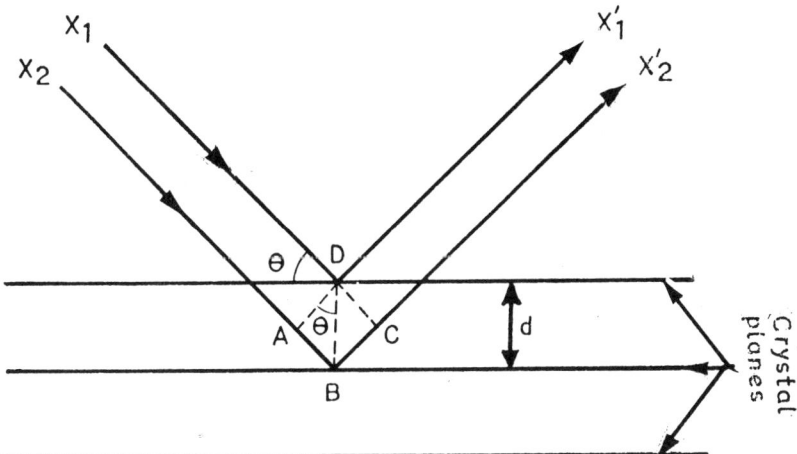

Fig.6.7: Diffraction of x-rays by an analysing crystal

(6.5) is known as Bragg's equation. For a particular crystal, d is constant and hence the wavelength (λ) of x-rays is proportional to sin θ. An important

consequence of this result is that, for a given glancing angle, θ, only x-rays of specific wavelength are diffracted by the crystal. If the x-ray beam consists of several wavelengths, they can be diffracted simply by changing the angle θ (e.g., by rotating the crystal). Such a crystal is known as an analysing crystal.

For precise measurements, the crystal should be as perfect as possible, so that the interplanar spacing (d) is constant in all parts of the crystal. For better results, instead of a plannar crystal , a curved crystal (shown in Fig.6.8) may be employed. This crystal serves as a dispersing device as well as a focussing device. Curved crystals are made of sodium chloride, lithium fluoride, quartz, germanium etc. The choice of material for the crystal depends on the wavelength range to be investigated (See tutorial 6.2).

(e) Detectors

For qualitative analysis, photographic films may be used for recording the fluorescence spectrum; but for quantitative measurements, their use is limited. The reason for this limitation is that the intensity of a spectral line (which is proportional to the concentration of the fluorescing element), recorded on a photographic film, is dependent on many variables, such as the

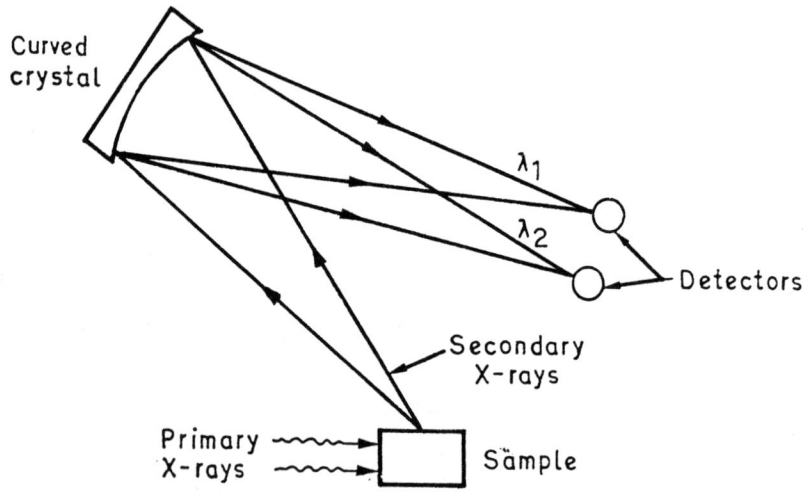

Fig.6.8 : Curved crystal optics

exposure time, processing of the film, type of emulsion on the film and so on; which are not under full control. Therefore, photoelectric means are, generally, employed for quantitative analysis. The basis of photoelectric detection of x-rays is discussed, in brief, as follows.

Consider a conducting cylindrical chamber filled with a gas of low atomic number e.g. helium. Assume, it contains a conducting wire placed along the axis of the cylinder and is insulated from it (as shown in Fig.6.9).

As the x-rays pass through the window, and strike the atoms of the gas; they get ionized. Consequently, primary ion-pairs (He = $He^+ + e^-$) are produced. If there is no potential difference between the central wire and the container, ionpairs recombine and no current flows in the external circuit; and if the potential difference exists, the electrons (e^-)are attracted by the central post and a current starts flowing. As the voltage on the central wire is increased, the current increases in a manner shown schematically in Fig.6.10.

Initially, as the voltage is slowly increased, the number of electrons reaching the central post increases; and hence the current also increases (as shown by part A in Fig.6.10). At a potential of V_a, all the electrons reach the central post and hence there is no increase of current with small increase in potential. This is shown by part B of the curve. This is the range in which the ionization chamber operates. With further increase in voltage beyond V_b, the electrons get accelerated. Some of these electrons acquire sufficient energy

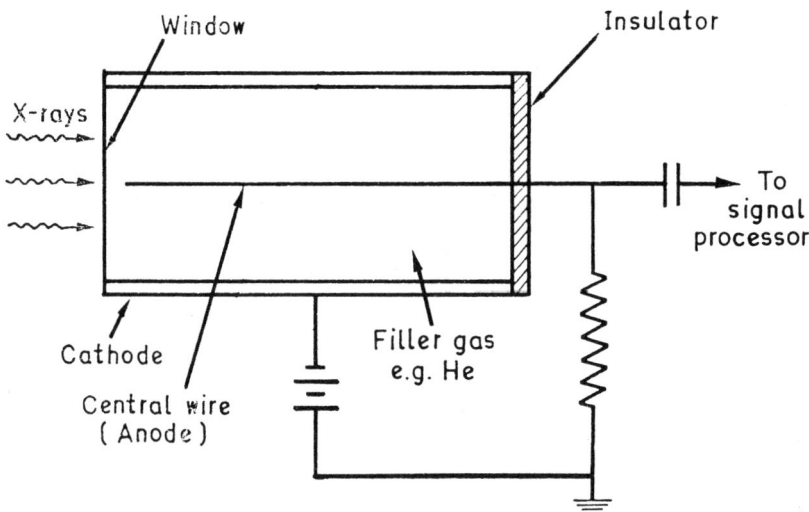

Fig.6.9 : Schematic diagram of a photoelectric detector of x-rays

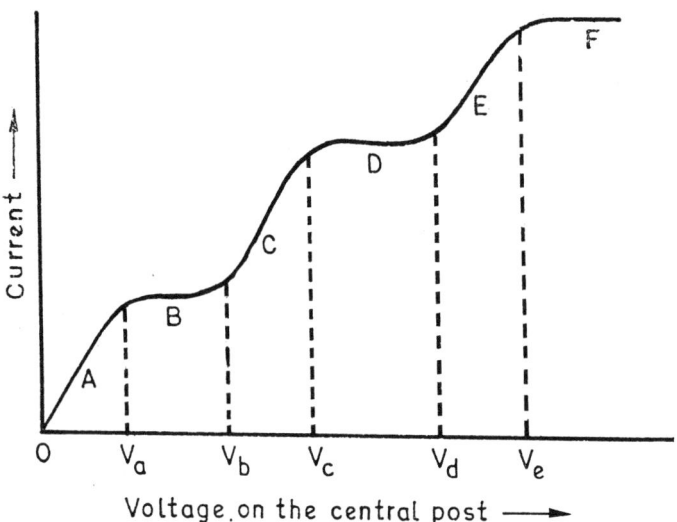

Fig. 6.10 : Variation of photoelectric current as a function of voltage on
the central wire. (Schematic).

and upon collision with the atoms of the gas create secondary ion-pairs,
thereby increasing the number of electrons. This causes more current to flow

in the external circuit and is shown by part C. It is not desirable to work in this range because slight change in potential causes drastic changes in the current. As the voltage reaches V_c, all the electrons (produced in the primary ion-pair formation) create secondary ion pairs. This gives rise to a stabilized condition (shown by part D). The current produced in this range is proportional to the number of ion-pairs formed and hence to the intensity of x-rays. This is the region in which a proportional counter operates. With further increase in voltage (i.e. , beyond V_d), the electrons produced in primary and secondary ion-pairs are accelerated so much that they produce more ion-pairs and this results in an avalanche of electrons reaching the central wire. The increase in current caused by these electrons is shown by part E. The part F shows that the response of the detector abeyond V_e is stable; however the current is dependent on the intensity of x-rays falling on the detector. This is the basis for the operation of Geiger Counter. These detectors are sensitive at longer wavelengths.

For shorter wavelengths another class of detectors known as scintillation counters are employed. In a scintillation detector, the x-rays are made to fall on a suitable substance (solid or liquid form) that absorbs this radiation and, in turn , emits visible or u.v. radiation; (the phenomenon being known as scintillation). The radiation emitted by this substance is then detected by a photomultiplier tube which is sensitive to uv and visible (but not to x-rays). The schematic diagram of solid and liquid scientillation counters are shown in Fig.6.11 (a & b). The compounds used for scintillation include thallium activated sodium iodide, anthracene, p-terphenyl in xylene, and naphthalene.

6.3.3. A typical configuration

The configuration of a typical x-ray fluorescence spectrometer is shown in Fig.6.12 (a&b). The source supplying the primary x-rays is a high vacuum discharge tube fitted with a copper or molybdenum target (although targets of other materials may also be used for special purposes). A highly stablized high voltage of the order of 50 KV is obtained with the help of the modules, voltage stabilizer and a high voltage generator. The current stabilization is done by monitoring the d.c. x-ray tube current and also controlling the filament voltage. For removing the unecessary target lines, a primary beam filter is, generally, used.

The primary x-rays irradiate the sample which is held in the sample holder. The latter is usually spun during irradiation, to minimize the surface effects. Commercial designs provide a sample chamber which can accomodate several samples on a turntable to which is fitted an automatic sample changer.

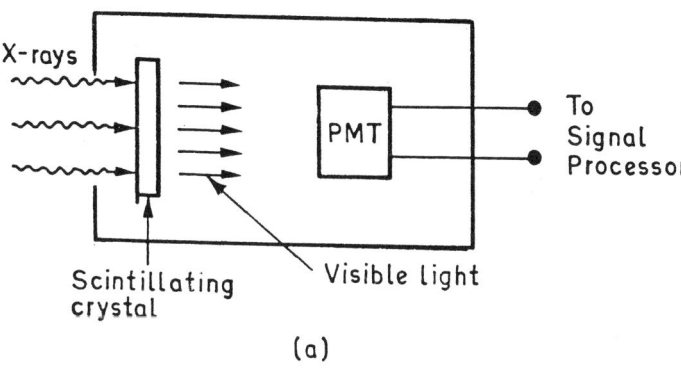

(a)

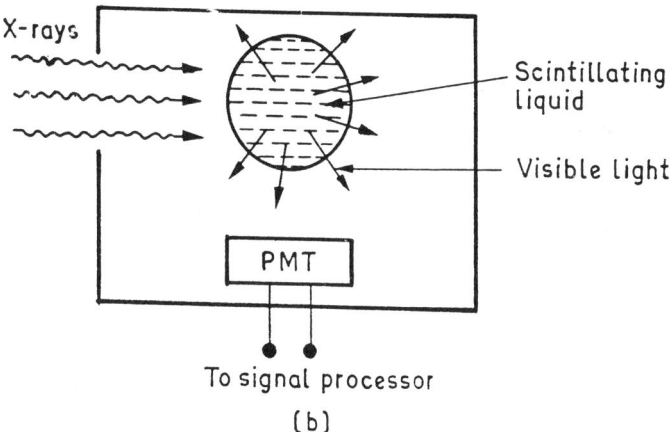

To signal processor

(b)

Fig.6.11: Schematic diagram of
(a) Solid state scintillation counter, and
(b) liquid scintillation counter.

The secondary x-rays are emitted by the sample in all directions and hence slits are required to collimate these rays and direct them onto the analysing crystal, as shown in Fig.6.12(b). In order to increase the range of wavelengths a multiposition crystal changer is usually incorporated. This may include upto 6 crystals arranged on a turntable. For scanning the emission spectrum in the desired range an appropriate crystal is rotated by a motor. A practical limit for the variation of θ is from 4°(minimum) to 75°(maximum). The secondary x-rays that are diffracted by the analysing crystal are allowed to pass through a second collimator to enter the detector. Normally, there is a choice of two detectors viz.

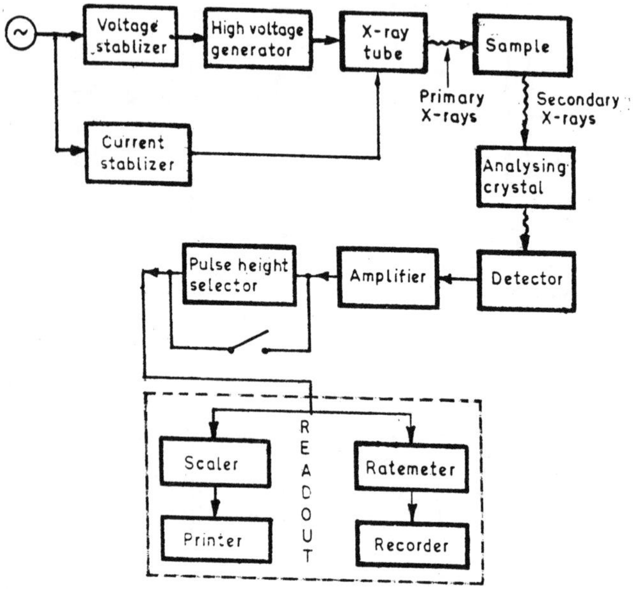

(a)

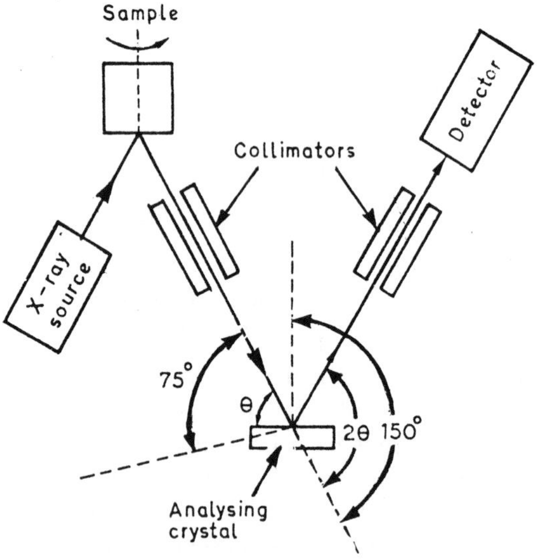

(b)

Fig.6.12 : A typical x-ray fluorescence spectrometer

(a) Block diagram showing different components
(b) Optical arrangement

(i)   a proportional counter for the detection of longer wavelengths and

(ii) a scintillation counter for detecting shorter wavelengths.

It can be visualized from Fig.6.12(b) that the detector has to be rotated at twice the angular speed of the analysing crystal.

The detector absorbs the diffracted secondary x-rays and converts them into pulses of electric current. The amplitude of each pulse is proportional to the energy of x-ray photon incident on the detector. These pulses are then amplified by an amplifier and selected by a pulse-height selector.

Why do we need a pulse height selector? We know, from eqn 6.5, that at a particular glancing angle, θ, an analysing crystal may diffract more than one wavelength in different orders. Since each wavelength will produce a pulse of different amplitude, some device is needed which can sort out these pulses and transmit only those having a pre-selected height. This function is performed by a pulse height selector. In this context, it may be called an electronic order sorting device. If it is not required, it may be bypassed.

The readout consists of either the ratemeter coupled to the chart recorder (for qualitative analysis) or scaler connected to the printer (for quantitative measurements) or sometimes, both. The ratemeter integrates the amplified pulses and displays them on a chart recorder. The function of scaler-timer system is to count the pulses in terms of the number of counts per preset time and record them on a printer.

6.4 X-RAY DIFFRACTOMETER

Rearranging eqn (6.5), we see that

$$d = \frac{m\lambda}{2 \sin \theta}$$

Thus, knowing the angle of diffracion, θ, it is possible to deduce the interplanar spacing, d, of the crystal. If θ is observed for different sides of the crystal, it is possible to determine the lattice of a crystal. Thus employing a monochromatic beam of x-rays (wave-length, λ), it is possible to know the size and shape of the unit cell of a crystal and hence the structure of the crystal.

The equipment needed for x-ray diffraction analysis, normally, takes the shape of a cylindrical camera. The design of this camera is based on the reciprocal lattice concept that is explained in the next subsection.

6.4.1. Reciprocal lattice concept

According to this concept, each plane represented by the Miller indices(hkl) in a crystal may be represented by its normal drawn from a common origin.

The length of the normal is taken to be inversely proportional to, the interplanar spacing, d_{hkl}. Thus the length and direction of the normal uniquely describe a set of parallel planes. It has been found that the terminal points of all such possible normals (drawn from a common origin) form a lattice array. This is known as the reciprocal lattice.

How can such a lattice be constructed ? Consider an unit cell of a typical crystal shown in Fig.6.13 (a).

If we sight along b-axis from the upper face ABCD, different planes (100), (001), (101) and (102) inside the unit cell, will appear as shown in Fig. 6.13 (b). Accordingly, the cell edges in the plane of the Fig.6.13(b) are AB = a and BC = c. As all these four planes are parallel to b-axis, their normals will be in the plane of the figure. In order to locate the points representing these planes; (i) draw the normal to each plane from a common origin (say, B); and (ii) take the length of the normal to be equal to $1/d_{hkl}$, where d_{hkl} is the interplanar spacing for (hkl) planes. The terminal points of all such normals form a lattice array as shown by dashed lines in Fig.6.13(b). This is known as reciprocal lattice because the distance of each terminal point from the origin is reciprocal to the interplanar spacing of the planes that it represents. The terminal points are called reciprocal lattice points and the normals are known as reciprocal lattice vectors (σ_{hkl}).The magnitude of σ_{hkl} is $1/d_{hkl}$ and its direction is parallel to the normal to the (hkl) plane.

In order to see the application of this concept to x-ray diffraction, rewrite equation (6.5) as follows.

$$\sin \theta_{hkl} = \frac{\lambda/2}{d_{hkl}} = \frac{1/d_{hkl}}{(2/\lambda)} = \frac{\sigma_{hkl}}{(2/\lambda)} \qquad (6.6)$$

In eqn (6.6), we have taken $\theta = \theta_{hkl}$, $d = d_{hkl}$ and m=1. Let us now construct a circle (see Fig.6.14(a)) whose diameter AO = $2/\lambda$; draw a line OP = $1/d_{hkl}$ and join the points A and P by another line AP. The angle APO is obviously 90°. Take the angle OAP = θ_{hkl}. We can see that the relation (6.6) is easily verified by the geometry of Fig. 6.14(a). With this construction, the geometrical interpretation of x-ray diffraction may be given as follows.

Assume that the x-ray beam enters the circle at a point A, passes along the diameter AO, and leaves the circle at a point 0, if there is no obstruction. If we place a crystal at the centre, S, of the circle and orient it such that it is parallel to AP (see Fig.6.14 b). Since OP is normal to AP and hence it is also normal to the reflecting plane (say, hkl) of the crystal. Thus OP has the direction of the reciprocal lattice vector (hkl) and its length is also $1/d_{hkl}$. From Fig.6.14(b), it is also obvious that $\angle OSP = 2 \angle OAP = 2\theta$ and hence the direction of the vector from the centre, S, to the reciprocal lattice point, P_{hkl},

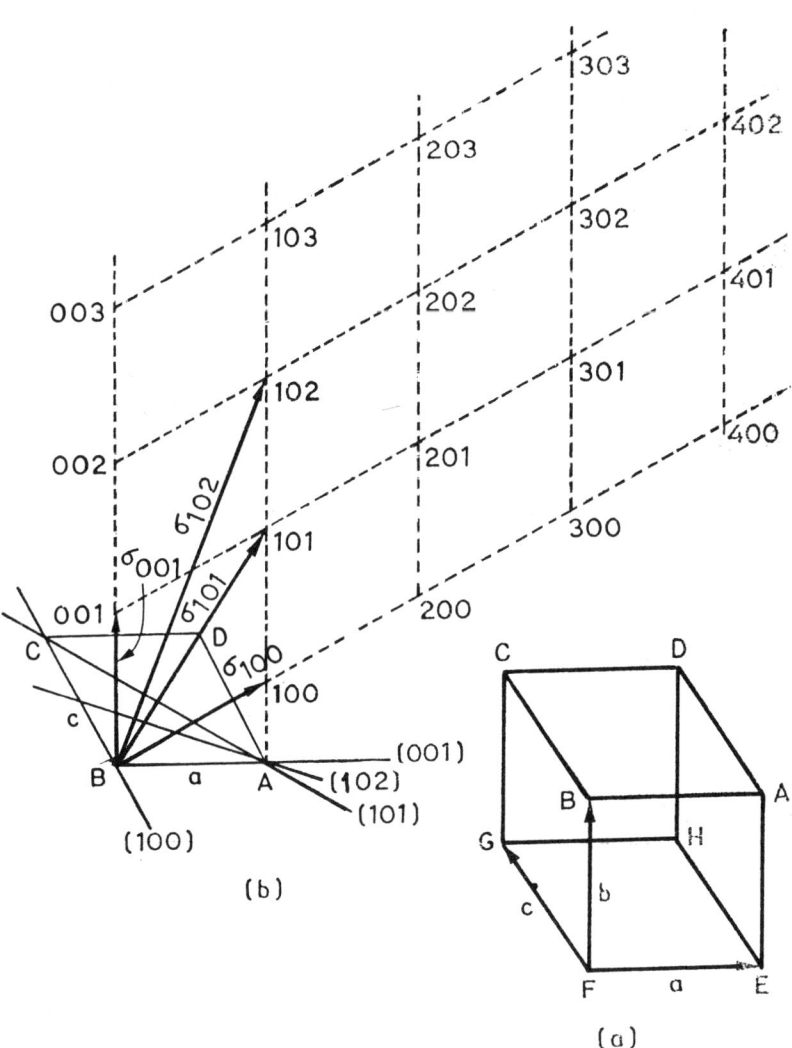

Fig.6.13 (a) Unit cell of a typical lattice, and
(b) Construction of a reciprocal lattice.

represents the direction of the x-ray reflection.

Thus, in a diffraction experiment, the crystal a can be pictured as located at the centre of a circle of radius $1/\lambda$ and the origin of the reciprocal lattice of the crystal is centred at a point where the direct beam leaves the circle (i.e., the point 0). Whenever the reflecting plane (hkl) of the crystal makes an angle θ with the direction of the direct x-ray beam, the terminal point (P_{hkl}) of σ_{hkl}

vector lies on the circumference of the circle and the reflected x-ray beam passes through this point. Therefore, the diffraction can occur only when a reciprocal lattice point (i.e. the terminal point) touches the circle. In three dimentions, the circle of Fig.6.14(b) becomes a sphere and is known as the sphere of reflection or Ewald Sphere (after Sir P.P. Ewald who developed this concept).

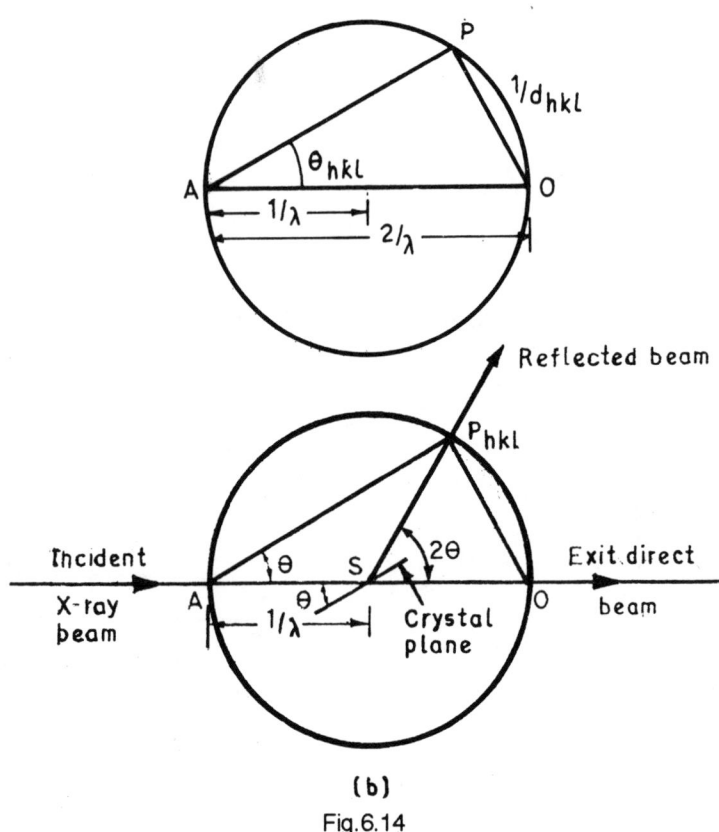

(b)

Fig.6.14

6.4.2 The powder method of x-ray diffraction

Most materials are not accessible in single crystal form; however, their powders are easily available. The powder of a crystalline material may be thought of as a random aggregate of tiny crystallites. The study of such materials is carried out by the powder method.

Referring to our previous discussion, what will happen if, instead of a single crystal, a powder is placed in the path of the x-ray beam ? Consider the same set of planes (hkl) in each crystallite of the powder. Since the crystallites are oriented randomly in different directions, these planes are also oriented randomly. Consequently, the reciprocal lattice vectors (σ_{hkl}) have all possible orientations. Thus the reciprocal lattice points(i.e. P_{hkl}) of these planes (hkl) form a sphere of radius, σ_{hkl}. The reciprocal lattice of a powder, therefore, consists of concentric spheres. The manner in which such a sphere intersects with the Ewald sphere is shown in Fig.6.15. This intersection occurs along a circle and the x-rays diffracted by the (hkl) planes in each crystallite form cones which are concentric with the direct x-ray beam. With reference to 6.14(b), it is clear that the half opening angle of the cones is 2θ. Two such cones are shown in Fig.6.15.

If photographic film is placed on the circumference of the circle in Fig.6.15, the film will be exposed in those parts where the diffracted cones intercept the film. The recorded photograph will appear as shown in Fig.6.16.

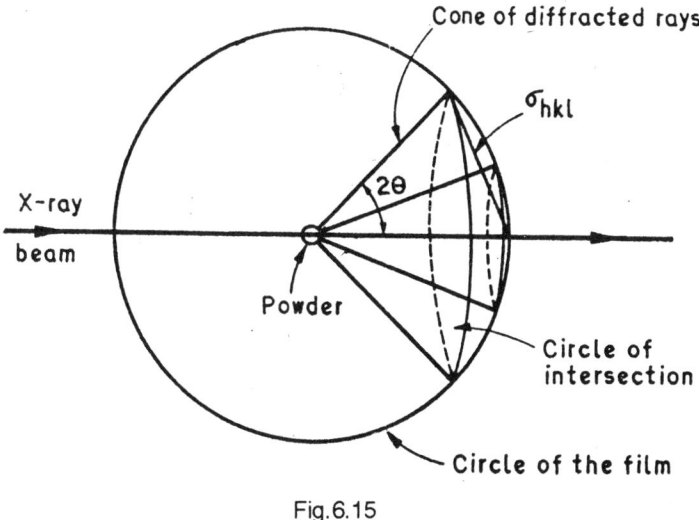

Fig.6.15

The glancing angle θ can be measured from this photograph if the sample to film distance is known. The necessary relationships are depicted in Fig 6.17. From this figure, it is clear that,

$$4\theta = \frac{S}{R} \quad \text{radians}$$

$$\text{or} \quad 4\theta = \frac{180\,S}{\pi R} \quad \text{deg} \tag{6.7}$$

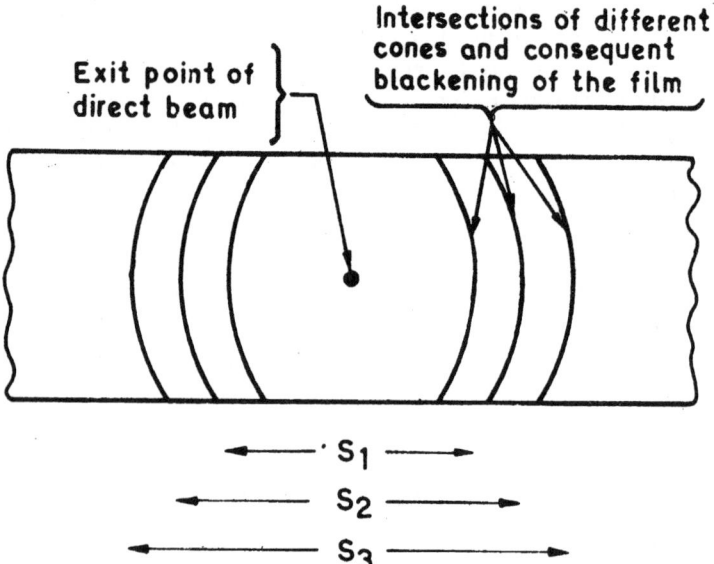

Fig.6.16 : Diffraction pattern of a powder

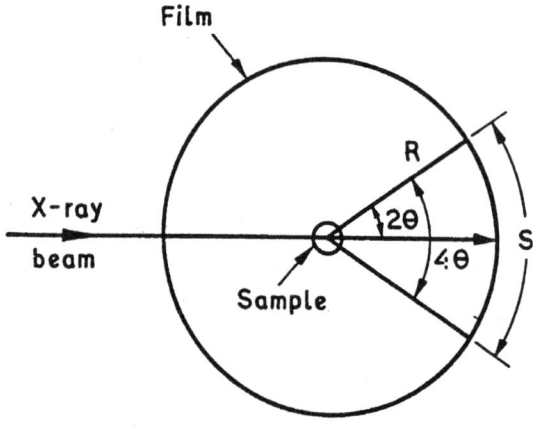

Fig.6.17

Thus the full opening angle of the diffraction cone 4θ, can be calculated by measuring the distance, S, between two corresponding arcs, recorded on the photograph, that are symmetrically displaced about the exit point of the direct beam. In eqn (6.7), R is sample (i.e. the powder)-to- film distance, usually the radius of the camera housing the film. This is called a Debye-Scherrer camera. Once θ is measured, eqn (6.6) can then be used to compute the d-values for each reflection.

In modern instruments, photoelectric detectors are used in place of photographic film. The mechanisms of rotation of the specimen and the detector are complex. The detector is coupled to a chart recorder or to a computer through appropriate signal processing stages.

6.5 X-RAY ABSORPTION SPECTROMETER

Each element has a characteristic x-ray absorption spectrum. Therefore, the wavelength, at which a sudden change in absorption occurs, can be used to identify the element present in a given sample. The intensity of transmitted beam can be related to the concentration of the analytical species (i.e the element) present in the sample.

The absorption of monochromatic x-rays (wavelength, λ) through a homogenous matter of density ρ and thickness x is given by the Beer's law (see Fig.6.18).

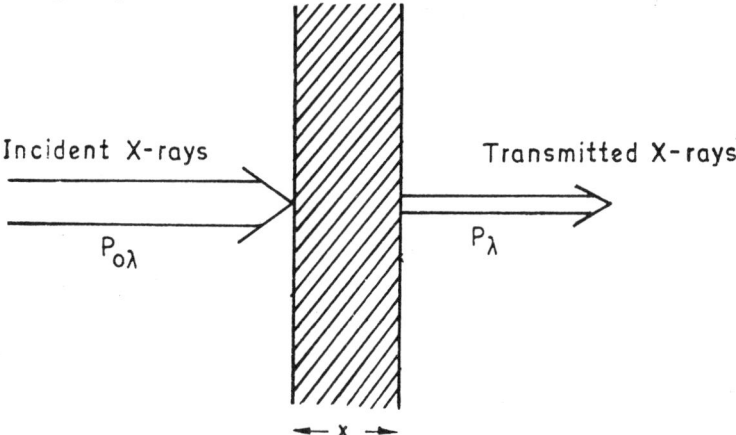

Incident X-rays   Transmitted X-rays

$P_{o\lambda}$ P_λ

← x →

Fig.6.18 : Schematic diagram showing absorption of x-rays by matter of thickness x.

$$P_\lambda = P_{o\lambda} \exp\left(-\mu_m \rho x\right) \qquad (6.8)$$

where P_λ and $P_{o\lambda}$ are respectively radiant powers of the transmitted and incident x-ray beams., μ_m is known as the mass absorption coefficient. It is dependent on the wavelength of x-rays and the absorbing species.

$$\mu_m = c\, z^4\, \lambda^3 \left(\frac{N_A}{A}\right) \qquad (6.9)$$

Here c is constant characteristic of the absorption edges, z is the atomic number of the element, λ is the wavelength of incident x-rays, N_A is the Avogadro's number and A is the atomic weight of the element. In a mixture, the mass absorption coefficient (μ_{mT}) is an additive function of the mass absorption coefficients of constituent elements.

$$\mu_{mT} = \mu_{m1} \, w_1 + \mu_{m2} \, w_2 + \; \text{-} \; \text{-} \; \text{-} \; \text{-} \tag{6.10}.$$

where μ_{m1}, μ_{m2}, etc are mass absorption coefficients of elements 1,2 etc and w_1, w_2 etc are their weight fractions.

What should be the design criteria for the x-ray absorption spectrometer?

First of all, a source of x-ray radiation followed by a wavelength isolation device would be needed, as usual. X-rays coming out of the monochromator should then strike the sample (or reference) which will either absorb or transmit them. Then there should be some arrangement for detecting and recording the intensity of the transmitted beam. Accordingly, various modules of the instrument may be arranged as shown in Fig.6.19.

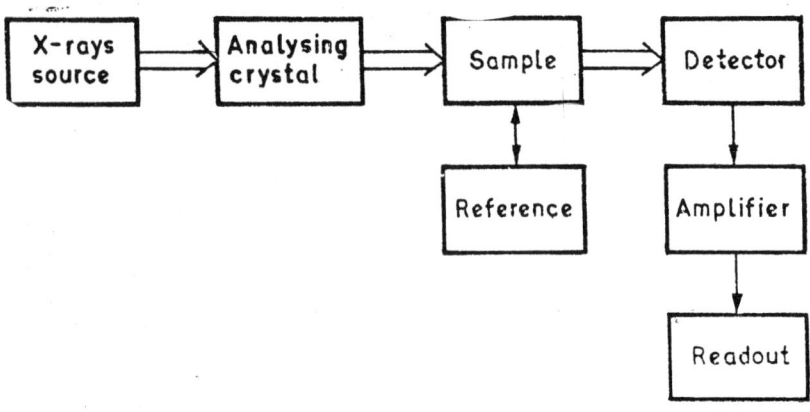

Fig.6.19 : Block diagram of a generalized x-ray absortpion spectrometer.

The signal(S) developed by the detector can be related to the modular performance as follows:

$$S = [P(\lambda). \, M(\theta , \lambda). \, B(\theta, \lambda)] \exp (-\mu_m \rho x) \, D(\lambda) \tag{6.11}$$

The factors within the first bracket are simply those collected in the eqn.(6.8) as $P_{o\lambda}$, the radiant power transmitted by the blank (or reference). Thus $P(\theta, \lambda)$ is the power furnished by the x-ray source as a function of wavelength (λ), $M(\theta, \lambda)$ is a factor that governs the solid angle (θ) as seen by the monochromator (i.e. the analysing crystal) and its transmission efficiency with λ, $B(\theta, \lambda)$ is the transmission of the blank (e.g. air or the solvent) and $D(\lambda)$ is the response of the detector with wavelength λ. It is now desirable to rewrite the above relation as follows.

$$S = P_{o\lambda} \exp (-\mu_m \rho x) D(\lambda) \qquad (6.12)$$

Thus, for reliable measurements, the components should be chosen such that the signal, S, is large enough for processing by the following modules. The only limitation with such a device would be that x-rays with continuously variable wavelength may not be available because every target element emits its characteristic spectrum. However, this problem can be solved to some extent, if more than one targets and more than one recording channels are employed.

6.6 ANALYTICAL APPLICATIONS

The x-ray fluorescence spectrometer is primarily employed for elemental analysis. All the elements with atomic number greater than 12 can be analysed by this instrument. With special equipment (which employs an evacuated optical system), elements with atomic numbers 5 to 11 can also be analysed. An important feature of the fluorescence method is that the intensity of x-rays emitted by the element (present in the sample) is independent of the chemical state of the sample. Consequently, chemical preparation of the sample prior to fluorescence measurements is, generally, not required. Furthermore, the method is nondestructive in nature and hence the sample can be reused after the analysis. The fluorescence spectra of elements are relatively simple and the overlap of x-ray emission lines of different elements is frequently uncommon. However, the line intensity should be corrected for the background emission which results from the general scattering of x-rays.

The qualitative analysis (i.e. the detection of elements in a sample) can be performed by measuring the angle of diffraction, θ of fluorescent x-rays. This measurement helps in computing the wavelength of fluorescent x-rays

$$\left(\lambda = \frac{2d}{m} \sin \theta \right)$$

Since each element emits its characteristic wavelength (λ), its knowledge helps in identifying the fluorescing element.

In order to perform quantitative analysis ,(i.e. to measure the concentration of the fluorescing element), the intensity of the characteristic x-ray emission line of the element is determined as follows. For a particular glancing angle θ (i.e. the angle between the primary x-ray beam and the surface of the planar analysing crystal), the detector is set at an angle of 2θ with respect to the direction of the primary beam and the counts are collected for a fixed period of time. The number of counts per second, cps; (say, a) gives the intensity of the emission line plus the contribution to intensity due to the background. The detector is then set at a nearby portion of the spectrum (where no emission line lies), and the counts are again collected for the same time. The number of counts per second (say, b) gives the intensity contribution due to background. The true peak intensity of the x-ray emission line is, therefore, I = a - b. The value of I is then related to the concentration of the fluorescing element via a calibration curve.

The applications of x-ray diffraction and absorption techniques are discussed in Tutorial 6.3 and 6.5 respectively.

TUTORIAL-6

6.1 When the energy of the exciting x-rays (i.e. primary x-rays) just equals the energy required to remove an electron from the inner shell of the atom, the exciting radiation is strongly absorbed; that is, there is a sharp rise in the absorption of the exciting radiation. The wavelength or the energy corresponding to this limiting radiation is known as an absorption edge. The short wavelength cut off (λ_{min}) of the x-ray source must be equal to or be shorter than the wavelength of the absorption edge. Thus, there is a critical potential which must be applied to the x-ray tube. This is known as critical tube voltage (or potential), V_{crit}. (It should not be confused with critical excitation potential, (measured in eV), which is the energy of the x-ray photon corresponding to the energy of the absorption edge).

Thus, in the limiting case,

λ_{min} = λ of the absorption edge.

From relation (6.1), $\lambda_{min} = \dfrac{12400}{V}$ (A°)

Hence $V_{crit} = \dfrac{12400}{\text{absorption edge } (\lambda)}$

Using this formula, calculate the crtitical tube voltage (in KV) for exciting the K-lines of the following elements.

Element	$K_{edge}(A°)$
Cu	1.380
Mo	0.620
W	0.178

Answer: 8.985, 20.000, 69.663 KV

6.2 Calculate the useful range (in A°) of the following analysing crystals, with the d-values of the reflecting planes mentioned against them. Assume that θ can be varied from $4°$ to $75°$.

Analysing crystal	d-value (A°)
(i) LiF	2.014
(ii) NaCl	2.821
(iii) CaF$_2$	3.160

Hint: Useful range may be found from the following relations:

$$m \lambda_1 = 2d \sin \theta_1 \text{ and } m \lambda_h = 2d \sin \theta_h$$

Here λ_1 and λ_h are the lower and higher limits respectively of wavelengths that can be studied with a crystal of interplanar spacing, d. $\theta_1 = 4°$ and $\theta_h = 75°$ are respectively the lowest and highest values of the glancing angle, θ. Take m = 1.

Answers: (i) 0.281 to 3.89A°, (ii) 0.393 to 5.45A° (ii) 0.441 to 6.10 A°

6.3 Indexing of cubic crystals

We know that the interplanar spacing, d_{hkl}, can be related to the Miller indices (h,k,l). Thus for cubic crystals,

$$d_{hkl} = \left\{ \frac{1}{\sqrt{(h^2 + k^2 + l^2)}} \right\} a$$

where a is the edge of the cubic lattice (or the unit cell parameter).

Now, consider some families of planes (h,k,l) and note the corresponding values of $h^2 + k^2 + l^2$ (as given in Table below).

Plane represented by (hkl)	$h^2+k^2+l^2$
100	1
110	2
111	3
200	4
210	5
211	6
220	8
221	9
300	9
310	10
etc	etc

A close examination of this table reveals that the number 7 is missing. Whatever be the values of h, k, l, the sum of their squares will never be equal to 7. This is known as a forbidden number. Next two such numbers are 15 and 23. Another point of interest is that the sum of the squares of the indices 221 and 300 both equal 9. This simply means that the reflections from these planes overlap.

Now combining the expression for d_{hkl} for cubic crystals with the Bragg's equation $m\lambda = 2d_{hkl} \sin \theta_{hkl}$ and putting m = 1, we get

$$\text{Sin}^2 \theta_{hkl} = \frac{\lambda^2}{4a^2} \quad (h^2+k^2+l^2)$$

For a particular powder pattern, λ and a are constant and hence $\sin^2 \theta_{hkl}$ is proportional to $(h^2+k^2+l^2)$.

Thus, the above relationship may be used to recognise the powder pattern of the cubic lattice.

The procedure is as follows. Measure the position of the diffraction lines on the photographic film. Calculate the values of θ and write them in the increasing order and then calculate $\text{Sin}^2\theta$ values. Divide all these values of $\sin^2\theta$ by the lowest value of $\sin^2\theta$. If these ratios are approximately 1,2,3,4,5,6,8 etc; then the diffraction pattern is that of a cubic crystal. Once the sequence of numbers is established, the Miller indices(hkl) can be assigned. This is called indexing of cubic crystals.

Question: A power pattern of CsCl was recorded using Cu-K_α radiation (= 1.542 A°).

The observed Bragg angles are θ = 10.72°, 15.31°, 18.88°,20.91°, 24.69°, and 27.24°. Employing the above scheme, find the Miller indices (hkl) of the reflecting planes and the corresponding values of d_{hkl}. Also compute the unit cell parameter.

Answer: The Miller indices are 100,110,111,200,210,211 and the corresponding d-values are 4.14, 2.91, 2.37,2.15,1.84, and 1.68A° The unit cell parameter is 4.14 A°.

6.4. The diffraction pattern of a crystalline powder (structure: cubic) was recorded using a Debye-Scherrer camera of radius 57.3 mm. The x-ray tube was fitted with nickel target (with cobalt filter) so that K_α radiation of Ni (= 1.66 A°) was allowed to enter the camera. The distances between the corresponding arcs of the three first order lines observed on the developed film were measured as 77.5, 109.1, and 130.4 mm. Calculate

(i) the Miller indices of the reflecting planes,

(ii) the spacing, d_{hkl}, of the crystal in A°,and

(iii) the unit cell parameter (in A°).

Answer: (i) 100,110,111, (ii) 2.50,1.77,1.44, (iii) 2.5A°.

6.5. How can x-rays be employed for monitoring the thickness of metal sheets?

Hint: You may think of proposing a design of an x-ray absorption meter for this purpose.

7

NUCLEAR MAGNETIC RESONANCE SPECTROMETERS

7.1 INTRODUCTION

Nuclear magnetic resonance (abbreviated as NMR) spectrometry is based on the measurement of absorption of radiation in the radiofrequency range by a magnetic nucleus (such as 19 F) when subjected to an external magnetic field. The frequency range of absorption varies from about 4 to 600 MHz which corresponds to a wavelength range of about 75 to 0.5 meters. The absorption of radiation of any frequency corresponds to a transition between a pair of magnetic states or energy levels of the nucleus. Unlike atomic absorption which characterizes separate atoms, NMR transitions characterize atoms within molecules. Since each nucleus is shielded somewhat from the external magnetic field by the surrounding electrons, the frequency of an NMR transition varies with the environment surrounding the nucleus. The set of such absorption frequencies for one kind of magnetic nucleus in a sample is a NMR spectrum. If the external magnetic field is held constant the independent variable is the frequency and if the frequency is kept fixed, the magnetic field strength becomes the independent variable. NMR spectrometry is one of the most powerful tools for indentifying molecules and detailing their structure. A quantitative estimate of the absorbing species can also be performed employing this tedchnique.

7.2 DESIGN CRITERIA

Before we arrive at a particular criteria for designing a NMR spectrometer, it is essential to understand the fundamental aspects of nuclear transitions. It has been revealed by the nuclear studies that protons and neutrons comprising a nucleus have characteristic motions and spins similar to that of electrons in an atom. As a consequence, most of the nuclei possess an angular momentum*

---

*A more accurate value of angular momentum results on the application of Schroedinger equation. This is given by $[I(I+1)]^{1/2}$ h.

ħ I, where ħ = h/2π (h is the planck's constant) and I is the spin quantum number. Since a nucleus bears a charge, its spin gives rise to a magnetic field. The resulting magnetic dipole of moment, say, μ, is directed along the axis of spin.

For a proton, it has been shown that the magnetic dipole moment, μ, is related to the angular momentum by the following relations:

$$\mu = g \left(\frac{e}{2MC} \right) ħ I \qquad (7.1a)$$

or $\qquad \mu = \gamma ħ I \qquad (7.1b)$

where $\gamma = g(e/2MC)$ is called the magneto gyric ratio; e is the charge on a proton, M its mass and C is the speed of light; g is termed a nuclear g -factor.

However, identical values of allowed energy result from two treatments. The nuclear magnetic moment is generally reported in terms of the nuclear magneton, β_n where $\beta_n = e ħ/2MC$ and is equal to 5.051×10^{-31} JG^{-1}. Thus the value of μ for a proton is 2.7927 nuclear magnetons.

Energy levels in a magnetic field

In the absence of any external magnetic field, the energy attributable to the magnetic moment remains essentially constant, i.e., it does not change with orientation of the dipole. However, upon application of an external magnetic field, of strength, say, H_o, the nucleus is acted upon by a torque which tends to align the dipole parallel to the direction of the field. The classical treatment shows that the potential energy of such a system is given by the expression

$$E = - \mu H_o \cos \theta = - \mu_z H_o \qquad (7.2)$$

where the field H_o is along the positive z-direction and makes an angle θ with the magnetic moment μ. μ_z is the component of μ in the z direction.

Classically, the energy of such a system can assume infinite number of values depending upon the orientation of dipole with respect to the field direction. Quantum mechanics, however, limits the number of possible energy levels to a few. For a given spin quantum number I, there are $2I + 1$ magnetic quantum states, m, given by

$$m = I, I - 1, I - 2, \ldots\ldots -I. \qquad (7.3)$$

If these numbers m are substituted for I in eqn. (7.1b), we get the allowed values of μ_z as $\gamma \hbar I$, $\gamma \hbar (I - 1)$, ... $- \gamma \hbar I$.

In turn, the values of μ_z can be inserted in eqn (7.2) to get the allowed values of nuclear magnetic energy states as

$$E = - m \gamma \hbar H_o \qquad (7.4)$$

where m is given by eqn. (7.3). These levels correspond to different orientations of the nuclear magnetic moment with the field. Alignment in the direction of the magnetic field (positive value of m) corresponds to a lower energy and an increasingly antiparallel alignment (negative value of m) corresponds to successively higher energy levels.

Given a set of energy levels, what are the allowed energy transitions? The selection rule for transition among the nuclear magnetic energy levels states that for an allowed transition, $\Delta m = \pm 1$. Thus the energy of a transition for a proton is always

$$\Delta E = \gamma \hbar H_o \qquad (7.5)$$

because for a proton, for which, $I = 1/2$, the nuclear magnetic quantum numbers m are $+ 1/2$ and $-1/2$. The two energy levels correspond to the two possible orientations of the spin axis with respect to the magnetic field as shown in Fig.7.1.

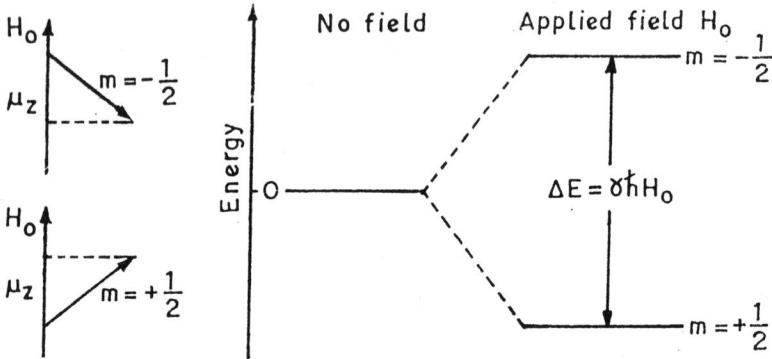

Fig. 7.1 : Magnetic moments and energy levels for a nucleus with a spin quantum number of 1/2. A transition from a lower level (m = + 1/2) to higher level (m= -1/2) requires an energy of transition $\Delta E = \gamma \hbar . H_o$.

The magnetic nucleus may be stimulated for an upward transition; if it is irradiated with electromagnetic energy of frequency f_o such that $hf_o = E$.

Since the angular frequency $\omega_o = 2\pi f_o$ we have

$$E = \frac{h}{2\pi} \; (2\pi\, f_o) = \hbar\, \omega_o \qquad (7.6)$$

Comparing eqns (7.5) and (7.6), we get

$$\Delta E = \gamma\, \hbar\, H_o = \hbar\, \omega_o$$

or $\qquad \omega_o = \gamma\, H_o \qquad (7.7)$

Eqn (7.7) is known as Larmor's equation and is the basis for NMR spectrometry.

Nuclear magnetic resonance

In the presence of an external field H_o, the nuclear dipole μ, experiences a torque that causes the dipole to precess about the direction of H_o at a rate given by

$$\frac{d\vec{\mu}}{dt} = \gamma\, (\vec{\mu} \times \vec{H}_o) \qquad (7.8)$$

The precession frequency is independent of the energy state and can be seen to be identical with ω_o given by eqn. (7.7).

If in addition to the steady magnetic field H_o acting along the z direction a radiofrequency field H_1 oscillating along y axis is applied; the nuclear dipole will experience a second torque.

The two fields will add to give an effective field H_e as shown in Fig.7.2. The rf field may be thought of (and is also shown) as the resultant of two superposed circularly varying fields of equal amplitude; one rotating clock wise and the other anticlockwise. Only that component will interact with μ which is rotating in the same sense. The other component will not interact and may be disregarded. As the frequency of the rf field is increased gradually there is increasing interaction of H_1 with the precessing magnetic moment. When the angular frequency of H_1 ($2\pi f$) equals the Larmor frequency ω_o, resonance absorption will occur. This resonance is shown in Fig.7.2 as a simultaneous precession of μ about H_e and of H_e about H_o.

Once the system is raised to the excited state, sooner or later it will return to its lower energy state. The relaxation may occur by the emission of radiation of frequency corresponding to the energy difference between the states (fluorescence). However, radiation theory shows that this process can not occur to a considerable extent. Thus in nuclear magnetic resonance, the relaxation occurs primarily by non-radiative processes.

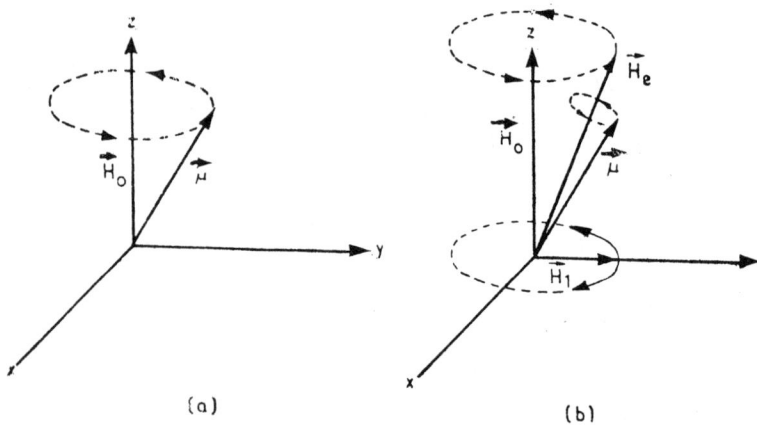

Fig.7.2 (a) The precession of the magnetic moment μ in the presence of external steady magnetic field, H_o, along z-axis;

(b) The precession of μ under the combined influence of both steady field H_o (along z-axis) and a rf magnetic field H_1 (along y axis).

The basic configuration

From an instrumental point of view, since the actual magnitude of any NMR absorption signal is minute, a high resolution NMR spectrometer would require at least the following modules:

(i) a stable magnet for producing a very steady, strong and homogenous field, H_o

(ii) a highly stable source of rf radiation, i.e. a rf oscillator (or transimitter), to produce a low power (of the order of milliwatts) rf magnetic field H_1, from which energy can be absorbed for magnetic transitions,

(iii) a highly sensitive rf detector- amplifier (or receiver) for detecting the resonance signals of the sample

(iv) a linear sweep circuitory fpr varying either the magnetic field H_o (field sweep) or rf frequency (frequency sweep) through the range of interest and

(v) a readout device for displaying or recording the NMR spectrum.

Accordingly, the basic NMR spectrometer may be represented block diagrammatically as shwon in Fig.7.3. The various modules are discussed in the next section.

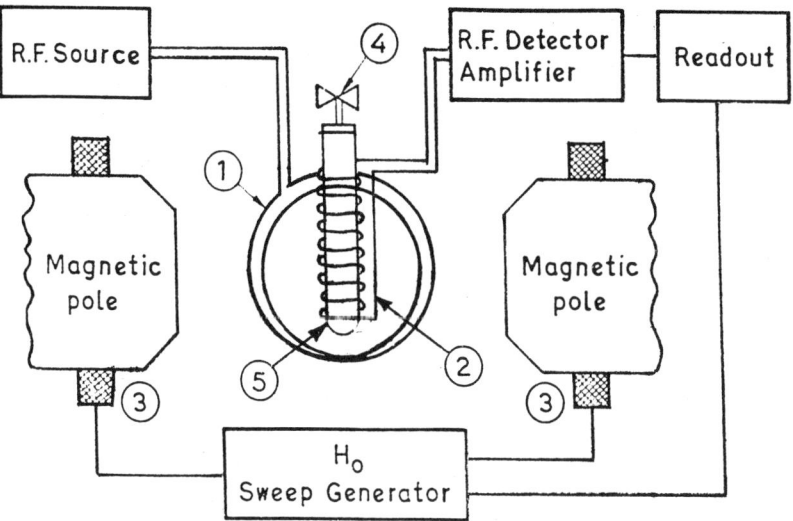

Fig. 7.3 : A schematic diagram of a NMR spectrometer. (1) Transmitter coil, (2) Receiver coil, (3) Field sweep coils, (4) Spinner and (5) Sample tube.

7.3 NMR SPECTROMETER: COMPONENTS

There are two types of high-resolution NMR spectrometers. They are

(i) continuous wave (cw) spectrometers and

(ii) Fourier transform spectrometers.

In the first case, the absorption signal is monitored continuously as the field (or frequency) is varied. In the latter, short, repetitive pulses of high-intensity, r.f. radiation excite all the nuclei simultaneously within a given frequency range. The output from the receiver is a time-domain spectrum that can be converted into a frequency domain by a Fourier transformation.

In the following subsections we shall discuss, in brief, the important components of a cw NMR spectrometer.

7.3.1 MAGNETS

In order to produce a steady magnetic field H_o, any of the following three types of magnets may be employed:

(i) permanent magnets,

(ii) electromagnets and

(iii) superconducting solenoids.

Given the requirements of homogeneity and time-stability, the practical limits on field strength-H_o for permanent magnets appear to be about 14 kilogauss (kG), electromagnets 23.5 kG and superconducting solenoids 100 kG or more. The proton resonance frequencies corresponding to these field strengths are approximately 60, 100 or 400 MHz. The advantages of different types of magnets are as follows. Permanent magnets are simple and free from maintenance problem, the electromagnets provide greater field strength without excessive complexity; and the superconducting magnets give an unmatched field strength. However, all types of magnets require shimming for reducing local inhomogeneities. This involves the use of homogenizing coils for imposing a small magnetic field of adjustable gradient. By current adjustment in these coils one can cancel the gradients present in the main field. Electromagnets and superconducting solenoids require a feedback stabilized d.c. power supply. Further all types of magnets require temperature control.

7.3.2 Radio-frequency source

Generally, a crystal-controlled rf oscillator is employed for obtaining stability and precision in frequency. Thus a crystal cut to oscillate at exactly 15 MHz (to 1 part in 10^7) may be used with two frequency-doubling amplifier circuits (which use harmonics of the crystal frequency) to obtain a 60 MHz output.

In order to scan the spectrum, a frequency sweeep can be provided by adding a variable small frequency to the basic frequency. For this purpose a second variable frequency oscillator is generally employed. For proton resonances, the frequency range of scanning from 600 Hz to 60 MHz is sufficient, while for nuclei like 19F and 11B the usual range of scan is from 16 KHz to 56.6 MHz. When the sweeep and basic rf frequencies are mixed both sum and difference frequencies (i.e. $\omega_o \pm \omega_m$) are obtained and hence a single side band modulator should be used to select the sum $\omega_o + \omega_m$ and reject the difference frequency.

7.3.3 Field sweep generator

If the radio frequency imposed on the sample is kept constant to obtain a NMR spectrum, the magnetic field H_o must be swept (as shown in Fig.7.3). In this case, the sweep coils (also known as Helmholtz coils) are placed either infront of or wound on the pole piece of the magnet. By changing the current in these coils the magnetic field experienced by the sample can be varied. In its simplest form, the sweep generator is a potentiometer whose wiper is driven over a dc voltage range and at a rate selected by the operator. Typically, H_o need be swept through 10 mG for a 60 MHz instrument.

7.3.4 Sample probe

It is a device for holding the sample contained in the sample tube in the narrow space between the magnetic pole faces. Besides acting as a sample holder, the probe has two more submodules. Firstly, it contains a transmitter coil connected to the radiofrequency generator. This coil is wound round the sample tube so that the sample is irradiated by the r.f. radiation of approipriate frequency. Secondly, the probe also has a receiver coil wrapped on the sample tube in a manner so as to be at right angles to the transmitter coil.

Theoretically, the sample shape should be an ellipsoid of revolution in order to minimize distortion of the magnetic flux. However, since most samples are in the liquid form, a cylindrical shape is a practical choice. The adequate length to diameter ratio of the cylindrical sample is of the order of 5.

Despite automatic shimming and use of a small sample, field inhomogeneities occur within the sample volume. If the axis of the rf coil and the sample tube is along y-axis, the inhomogeneities in the x and z- directions may be reduced considerably by spinning the sample tube at from 20 to 40 Hz. This averages out gradients of the order of 10^{-3} G. The desired spinning can be achieved by placing a air-driven rotor on the top of the sample tube.

7.3.5 Detector-Amplifier

In a single channel instrument, a single coil serves to produce the rf magnetic field and also detect the resonance absorptions. Some means must be provided for separating the two signals. One method employes an rf bridge in which the exciting signal is balanced against an equal amplitude reference with the modulation appearing as bridge unbalance and is extractable in that form. Where resonances are weak, this procedure yields poor response. In such cases, a crossed-coil arrangement (shown in Fig.7.3) is preferred. This employs separate coils for excitation and detection. If the steady field, H_o is directed along the z-axis and the sample tube and exciting rf coil are placed along the y axis, a detection coil may be placed along the x-axis. As the magnetic resonance produces a net magnetization in the x-y plane, the detector coil selectively picks up the resonance, while virtually excluding the rf field.

7.4 DUAL CHANNEL NMR SPECTROMETERS

In order to interpret a NMR spectrum, it becomes important to know accurately the relative separation of the resonance lines as well as their intensities in terms of the areas under the peaks. As the variables affecting such measurements are many, it is customary to determine the position of resonance lines of the sample with reference to a standard resonance peak. For proton NMR studies, tetra methyl silane, (TMS), $(CH_3)_4 Si$ is commonly used as a standard beacause its 12 magnetically equivalent protons give a single absorption line of unusual intensity. But this procedure solves only a part of the problem. Some instrumental means of calibrating the spectral scale must be employed.

In order to maintain the calibration of the instrument over a period of time, the design must provide for (i) the constancy of the ratio ω_o/H_o ie. the stability of the resonance condition, (ii) the stabilization of either field H_o or the frequency ω_o; and (iii) maintenance of the "baseline stabilization", i.e. the constancy of the oscillator output, detection as well as amplification stages. The double channel design, coupled with af modulation of H_o, is well suited to meet these requirements. The reference or control channel is locked on the sharp resonance line of a reference material, such as TMS. The sample channel is used to scan the spectrum of the sample. Both channels employ the same rf oscillator and the magnet. However, their rf coils, sample holders and detector amplifiers are separate.

In the double channel instruments, an audio frequency modulating field of frequency ω_m is also applied to both the sample and reference compound, to reduce the noise. As the modulation produces side bands on each line, ω_m is chosen to be greater than the width of the spectral range of the nucleus under study. In general, the modulation is applied to H_o. The af signal is imposed to separate coil in front of the pole faces of the magnet and its amplitude is regulated to give a oscillating field H_m of about 0.1 mG. A typical double channel NMR spectrometer featuring an af modulation is discussed in tutorial 7.4.

7.5. ANALYTICAL APPLICATIONS

The NMR spectrometer constitutes a powerful tool for the elucidation of the molecular structure. In most cases, 0.1 to 0.4 mg of the sample is needed. With a record of spectral information, as well as an empirical formula at hand, it is possible to examine the NMR spectrum and identify the substance.

Since the total area under a resonance line is directly proportional to the number of contributing nuclei, the NMR techniques are valuable in quantitative determinations. This technique is also helpful in elucidating the polymerization patterns of polymers. The NMR spectrometry has also contributed to the understanding of electron distribution in molecules, quantum mechanical nature of bonds and so on.

TUTORIAL-7

7.1 The sensitivity of cw NMR spectrometers is limited and hence it is not possible to obtain a good proton resonance spectra for materials that are available in micrograms. The commercial development of pulsed Fourier transform NMR spectrometers, has resulted in the vast improvement, in the sensitivity of NMR measurements.

In these instruments, the sample is irradiated periodically, by highly intense rf pulses of short duration (1 to 10 μ-sec) following which the free induction decay signal, a characteristic rf emission signal is recorded as a function of time.

What should be basic configuration of such instruments?

Hint: The magnets and the sample probe for a Fourier transform instrument are similar to those discussed in sec. 7.3 for cw instruments. A field sweep generator is unnecessary now. However, an additional digital computer may be required to control the pulses, accumulate the data and transform the measurement information for presentation on the CRT or the chart. A typical setup is shown schematically in Fig.7.4.

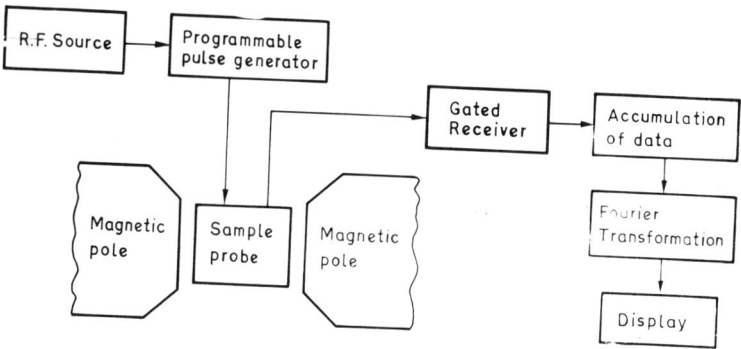

Fig.7.4. Block diagram of a typical Fourier transform NMR spectrometer

7.2. A typical cw NMR spectrometer employs a magnet that provides a magnetic field of strength 14.092 kG. At what frequency would the proton nucleus show resonance absorption?

Solution: The value of μ for a proton is 2.7927 nuclear magnetons, and one nuclear magneton equals 5.051×10^{-31} JG^{-1}. Thus

$$hf = \Delta E = \gamma \hbar H_o = \frac{\mu}{\hbar I} \hbar, \times \ H_o = \frac{\mu H_o}{I}$$

Therefore, $f = \dfrac{\mu H_o}{hI} = \dfrac{2.7927 \times 5.051 \times 10^{-31} \times 14092}{6.6256 \times 10^{-34} \times 1/2}$

$= 60 \times 10^6$ Hz

$= 60$ MHz.

7.3. (a) Which of the following two procedures would provide better signal to noise ratio ?

    (i) Single slow scan of an NMR absortpion peak.

    (ii) Averaging a set of faster scans of the peak requiring the same total time.

  (b) What conditions must hold for the above two procedures to yield comparable results?

7.4. The same type of nucleus, in different chemical environments(i.e. with different distribution of surrounding electrons) will experience a different effective magnetic field when put in a constant magnetic field (H_o) between the magnetic pole faces of a nmr spectrometer and hence there will be correspondingly slightly different resonant frequencies (for the same field, H_o). The effective field experienced by the nucleus is given by $H = H_o (1-\sigma)$

where σ is a dimensionless shielding constant, and may be either positive or negative.

    Keeping in view the above fact, what scale should be adopted for scanning the nmr spectra?

Hint: A widely used solution to this problem is that the position of the resonances of the sample is measured with respect to the resonance of the reference or standard material. For proton resonances, in nonaqueous media, a commonly used reference material is tetramethyl silane, $(CH_3)_4Si$, abbreviated as TMS, and its resonance position is assigned 0.00 on the δ - scale. The magnitude of the Chemical shift (δ) for the sample is expressed in parts per millilon (ppm) by the expression:

$$\delta = \left(\dfrac{H_{TMS} - H}{H_{TMS}} \right) \times 10^6$$

8

MASS SPECTROMETERS

8.1 INTRODUCTION

The mass spectrometer is an instrument that analyses the analytical species by converting it into a gaseous ionic form, followed by the separation of ions according to their mass-to-charge ratio. This technique is capable of providing qualitative as well as quantitative information concerning the molecular weight and the molecular structure of both the inorganic and organic compounds. It is one of the methods that can be employed to determine the molecular weight as high as 4000 and that is present in the sample as low as ppm level. Recently, mass spectrometry has been applied to the determination of atomic and molecular species on surfaces and for studying the compositional changes in solids as a function of depth. The principle of mass spectral measurements is simple but the equipment required is quite complex. Accordingly, the installation as well as the maintenance cost of the instrument is very high.

8.2 DESIGN CRITERIA

What should be the criteria for designing a mass spectrometer? In order to obtain a mass spectrum of the sample it must be converted into the charged particles or ions and hence the first requirement of the instrument is the source producing ions. As both positive and negative ions are produced, some means of separating one type of ions must also be provided in the same module. Next, a dispersion device commonlly called a mass analyser should be provided to isolate the ions of different masses. Finally, an ion detector followed by a signal processor stage and a readout device will be needed as usual. Accordingly, the modules of a mass-spectrometer may be arranged as shown block-diagrammatically in Fig.8.1. Unlike most optical spectrometers, the mass spectrometers need elaborate vacuum systems. The criteria for selecting the various modules and their configurations are briefly discussed in the following section.

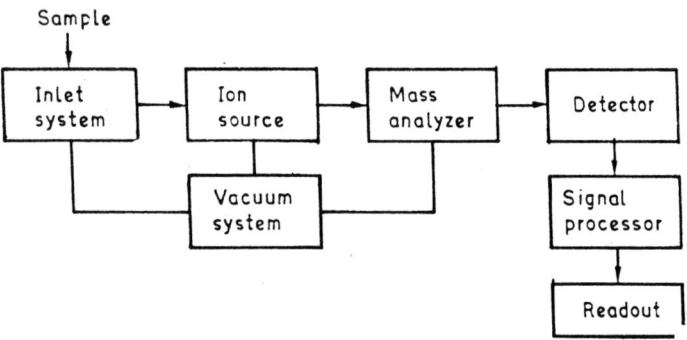

Fig.8.1 A schematic diagram of a mass spectrometer

8.3 COMPONENTS OF A MASS SPECTROMETER

8.3.1. Sample inlet system

This module is used to introduce the representative sample into the ion source. In order to accomodate different kinds of samples, three types of inlet systems are employed. There are :

   (i) batch inlet

   (ii) direct probe inlet and

   (iii) chromatographic inlets.

In the batch inlet system, the sample is volatilized externally and then allowed to leak into the evacuated ionization region. A schematic diagram of a system that can be employed for both the gaseous and liquid samples is shown in Fig.8.2. The introduction of gaseous samples involves merely a transfer of gas into the metering volume (about 3 ml) and then expansion into the reservoir flask (size about 3 liters)which is just ahead of the sample leak. The pressure in the metering volume ranges from 30 to 50 torr and that in the reservoir is 10^{-4} to10^{-1} torr. Liquid sample is introduced by touching a capillary or micropipette carrying the sample to a sintered glass disk that is covered with a layer of mercury or liquid gallium to prevent excess of air. The low pressure in the reservoir draws the liquid into the heating chamber and vaporises it immediately. Heating system extends the usefulness of the instrument to polar materials which are prone to be adsorbed on the walls of the chamber at room temperatures. From the sample reservoir, the vapor or the gas diffuse through a molecular leak into the ion source. The leak is a pinhole restriction (of diameter 0.005 to 0.02 mm) in a metal or glass diaphragm.

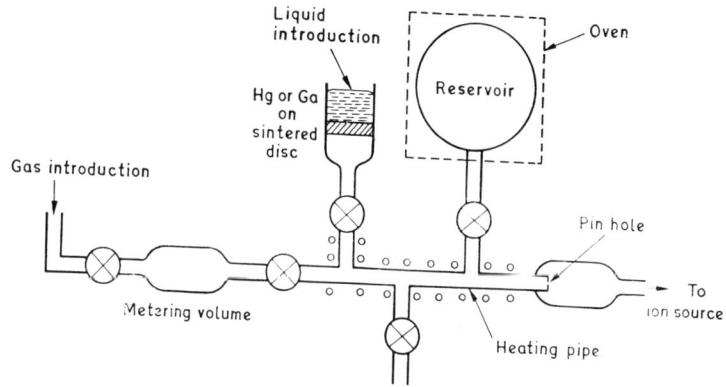

Fig.8.2 Schematic diagram of a batch inlet system

For introducing the non-volatile liquids, thermally unstable compounds and solids into the ionization region, a direct probe inlet system is employed. Generally, with such a system, the sample is loaded into a short length of a capillary, placed in the well end of the probe and positioned to within a few millimeters of the ion source through a vacuum lock. The temperature of the sample is then raised until the sufficient vapour pressure is indicated.

The effluent from a chromatograph column can also serve as a sample source. But the interfacing of the chromatograph with a mass spectrometer causes a problem because of the presence of the carrier gas. Generally, a jet separater is employed to isolate the sample from the carrier gas and feed it to the ion source.

8.3.2 Ion source

The ions for mass analysis can be produced by a variety of methods as given in Table 8.1. These methods can be grouped in two categories. In the first category known as gaseous ionization the ion source first volatilizes the sample following which the gaseous components are ionized by collision with electron or positive ions. In the second category, called desorption ionization, the bulk sample volatization is dispensed with. Instead, employing a direct sample probe, the energy is transferred (in a variety of ways) so that the sample in the condensed phase is transformed directly into a gaseous ionic form.

A standard method of producing ions, upon which the libraries of mass spectral data are based, consists of bombarding the gaseous or vaporized sample with a stream of energetic electrons. Fig.8.3 shows a schematic diagram of such a electron-ionization source or an ion gun. Here the electrons are produced by a heated filament (generally made of tungsten) which are then accelerated to about 70 eV by applying a potential difference between the filament and the anode.

Table 8.1: Ion Sources for Mass Spectrometers

Category	Method	Abbreviation	Ionizing agent
I	Electron ionization	EI	Energetic electrons
	Field ionization	FI	High potential electrode
	Chemical ionization	CI	Reagent positive ions
II	Field desorption	FD	High potential electrode
	Fast atom bombardmen	FAB	Energetic atoms
	Secondary ion mass-spectrometry	SIMS	Low fluxes of energetic electrons
	Plasma desorption	PD	High energy fission fragments
	Thermal desorption	TD	Heat
	Laser desorption	LD	Laser beam
	Electrohydrodynamic ionization	EHMS	High field

The energetic electrons strike the gaseous molecules, emitted from the molecular leak, and ionize some of them. The positive ions so produced are accelerated by applying a small potential difference between the repellers and the first accelerating slit. Further acceleration and collimation of the ions occurs when proper potential differences are imposed between the first accelerating slit and the focus and the second accelerating slit. This collimated and energetic ion beam then enters the mass analyser.

8.3.3. Mass analysers

This component of the mass spectrometer is analogous to an optical monochromater. The function of a mass analyser is to distingush between ions of different mass-to-charge ratio. Its capability to differentiate between two ions of masses m and m+ Δm is measured in terms of the resolution

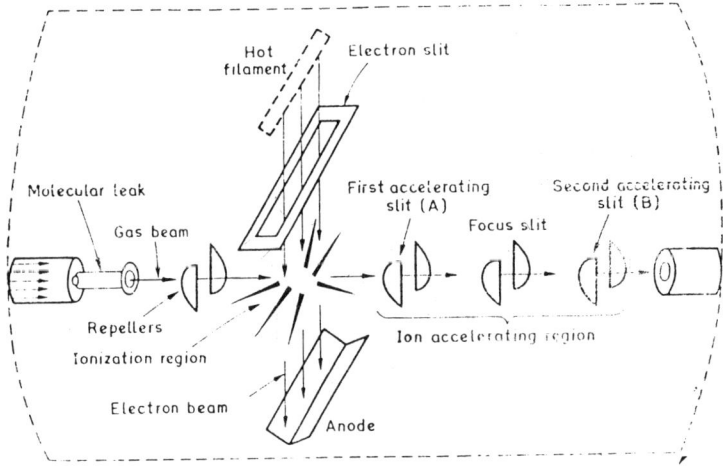

Fig.8.3 : Schematic diagram of an ion source

$m/\Delta m$. The resolution varies greatly with the design of the analyser. Two designs are discussed as follows.

(a) Single focussing magnetic sector analysers

Fig. 8.4 shows a schematic diagram of a typical 90° sector analyser. From the inlet system the gaseous sample is passed through the ion source, where the ions are formed by electron impact . The accelerated ion beam enters a metal analyser tube which is maintained at a pressure of about 10^{-7} torr. The circular section of the analyser tube is kept under the magnetic field which is provided by either a permanent magnet or an electromagnet. The ions of different mass can be focussed onto the exit slit of the analyser either by varying the magnetic field strength or by varying the accelerating potential between the slits A and B of the ion source.

If the potential difference between the slits A and B of the ion source is V, and if the ion is singly charged (charge e^+), then the kinetic energy of the ion of mass m which has acquired a velocity v at the entrance aperture of the analyser tube would be

$$E = eV = \tfrac{1}{2} mv^2 \tag{8.1}$$

As the ion enters the magnetic field of strength H, a centripetal force Hev acts on it. This is equivalent to mv^2/r where r is the radius of curvature of the path described by the ion. Thus, we have

$$\text{He v} = \frac{mv^2}{r}$$

$$\text{or} \qquad v = \frac{Her}{m} \qquad\qquad (8.2)$$

Substituting the value of v in eqn. (8.1) we get

$$eV = \tfrac{1}{2}m \left(\frac{H^2e^2r^2}{m^2} \right)$$

and the rearrangement of terms gives

$$\frac{m}{e} = \frac{H^2r^2}{2V} \qquad\qquad (8.3)$$

It follows from eqn. 8.3 that the mass spectra can be obtained by varying one of the three variables (H,r or V) while keeping the remaining two constant. Most of the modern sector type instruments, however, contain an electromagnet in which the ions are sorted out by varying the current in the magnet and thus H, and keeping r and V constant.

(b) Double focussing analysers

While deriving equation (8.3) it has been assumed that all the ions have the same kinetic energy at the entrance aperture of the magnetic sector. However, there is a small variation in the kinetic energy of ions of a given analytical species as they leave the ion source. This causes the broadening of the ion beam reaching the ion collector and hence a consequent loss of resolving power.

The resolving power of the analyser can be increased by first passing the ion beam through a radial electrostatic field as shown in fig.8.5. This field focusses the ions of same kinetic energy on slit 2, which then serves as a source for the magnetic separator. By this method, it is possible to resolve ions which differ by small fractions of a mass unit. Resolution of order of 2500 is easily achieved.

8.3.4 Detectors and signal processors

Three types of ion-detectors are commonly employed in mass spectrometers. They are :

    (i) faraday cup,

    (ii) electron multiplier and

    (iii) ion senstive photographic plate.

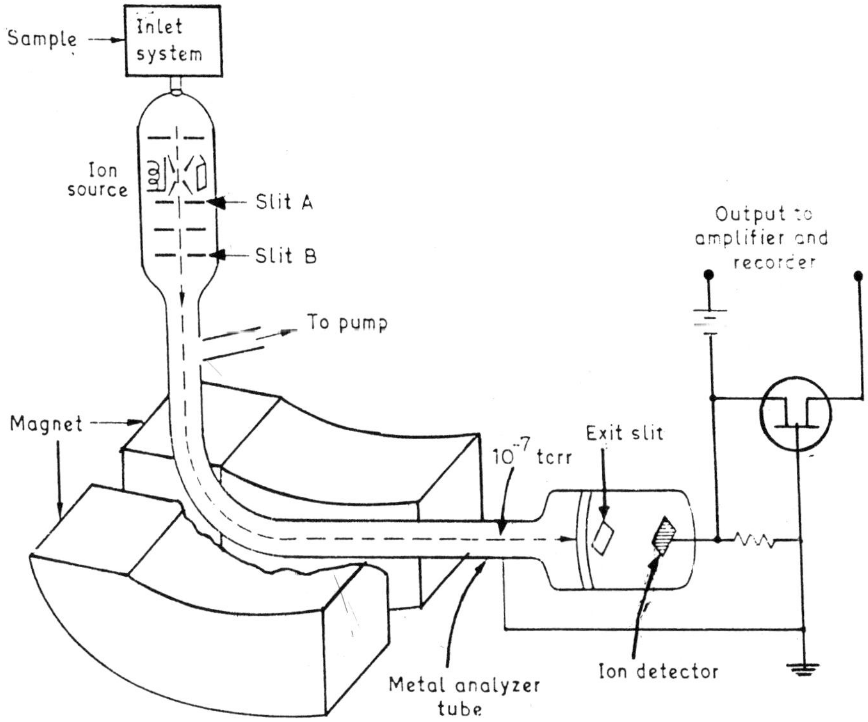

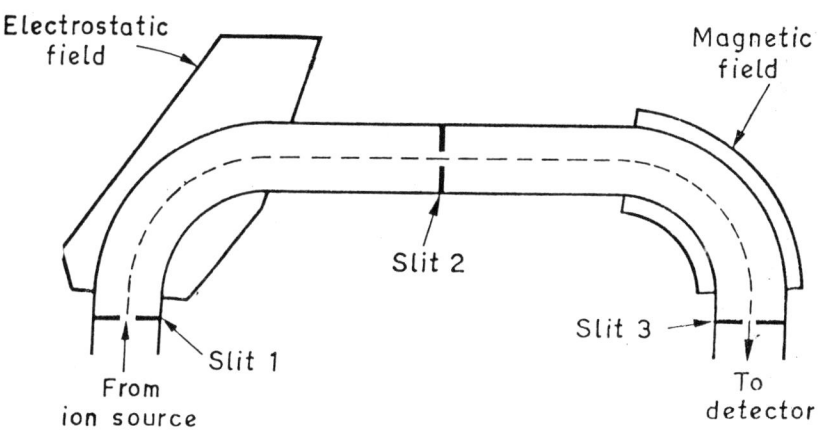

Fig.8 4 : Schematic diagram of mass spectrometer employing a single focussing magnetic sector analyser.

Fig.8.5 : Schematic diagram of a double focussing mass analyser

The Faraday cup detector consists of a small collector plate which is connected to the ground via a high ohmage resistor (as shown in Fig.8.4). The resulting potential drop across the resistor is impressed on a FET. The resulting potential is further amplified for display on the readout device. In an electron multiplier, the ion beam, from the exit slit of the analyser, is accelerated by applying a potential difference of 2 to 5 KeV. The accelerated beam strikes a plate which emits 2 to 3 electrons for each positive ion. The electron amplification is achieved via a secondary dynode arrangement.

The ion sensitive photographic plates are employed in Mattauch-Herzog type configuration. The plates are coated with AgBr in gelatin and are sensitive to positive ions.

The readout devices are analog or digital recorders, oscilloscopes, and computerized printers. Minicomputers and microprocessors have become an intergral part of most of the modern mass spectrometers.

8.4 SOME TYPICAL CONFIGURATIONS

A variety of mass spectrometric configurations is available. Some of these are discussed as follows. Fig.8.6 shows schematically the design of double focussing spark source mass spectrometer, which is based on Mattauch-Herzog geometry. In this configuration all the ions are focussed on a single focal plane regardless of m/e ratio. This arrangement is useful for photographic detection.

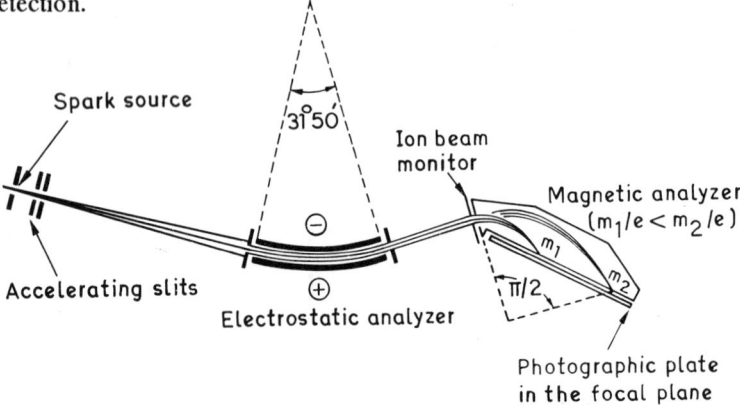

Fig.8.6 Schematic diagram of Mattauch-Herzog configuration

Fig.8.7 shows a quadrupole mass spectrometer that is compact, more rugged and less expensive than magnetic focussing instruments. Here the analyser consists of four short, parallel metal rods (of hyperbolic cross-section) arranged symmetrically around the ion beam, as shown. The opposite rods are electrically connected. One pair is attached to positive terminal of a variable dc source and the other pair to the negative terminal. A variable rf ac potential is also applied across the two pairs. None of these potentials accelerate the postive ions ejected from the ion source. However, the combined field causes the positive ions to oscillate about the central axis. Only those ions pass through the array which possess certain m/e ratio. Other ions are removed by collision with one of the rods. The mass spectrum is scanned by varying the frequency of ac supply while holding the potentials constant or by varying the potentials of two sources while keeping their ratio and also the frequency constant.

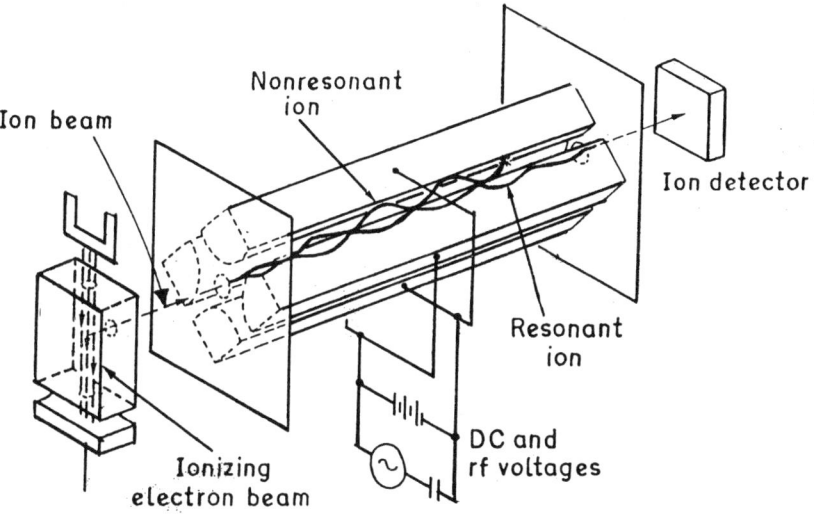

Fig.8.7 : A schematic diagram of a quardupole mass spectrometer

8.5 ANALYTICAL APPLICATIONS

Quantitative analytical applications of mass spectrometers fall in three groups. They are :

(i) the quantitataive determination of molecular species in organic and biological samples,

(ii) the quantitative elemental analysis of inorganic substances using spark sources and

(iii) the quantitative estimation of molecular or atomic species on solid surfaces.

The mass spectrometer is a powerful tool for qualitative analysis as well. In fact, initially it was developed for the study of isotopic abundance ratios and since then it continues to be the most important source for this kind of data

This data is useful for the determination of molecular formulae, dating of rocks and minerals, determination of isotope concentrations and so on.

Recently developed, secondary ion mass analysers and laser microprobe analysers are finding increasing use in determining the atomic and the molecular composition of solid surfaces.

Identification of unknown compounds is also possible. The mass spectral data gives the information about the molecular weight of the compound , its empirical formula, as well as the presence or absence of various functional groups. Final identification requires matching this data with spectral library of known compounds.

TUTORIAL-8

8.1 A typical single focussing magnetic sector mass analyser is operated at an accelerating voltage of 2500 V. In order to focus the peak for CH_4^+ ion on the detector, a magnetic field of strength 1250 G is required.

(a) Determine the range of magnetic field strength required to scan a mass range of 16 to 250, if the accelerating voltage is kept constant at 2500V.

(b) Determine the range of accelerating voltage required to scan a mass range of 16 to 250, if the magnetic field is kept constant at 1250 G.

Hint: Use eqn 8.3.

Answer: (a) 1250 to 4941G (b) 160 to 1250 volts.

8.2 If the mass analyser of tutorial 8.1 were operated at a constant field strength of 1250 G,

(a) what accelerating voltage would be required to focus the parent ion (M^+) for ephedrine (molecular weight 165) on the detector?

(b) what percent change in accelerating voltage would be required to scan from M^+ to $(M+1)^+$ ion for ephedrine?

Answer: (a) 242.4078 V (b) 0.6024%

8.3 **The ion cyclotron resonance spectrometer**

If a gaseous ion is introduced into a strong magnetic field, it starts circulating in a plane that is perpendicular to the direction of the field. The angular frequency ω_c of the motion of this particle can be obtained from eqn. 8.2. Thus

$$\frac{v}{r} = \omega_c = \frac{eH}{m} \tag{8.4}$$

where ω_c is the cyclotron frequency in rad/sec.

From this expression, it is clear that in a constant magnetic field, ω_c is inversely proportional to m/e ratio and is independent of the ion velocity. Increase in v causes a corresponding increase in radius. Such a circulating ion is capable of absorbing energy from a sinusoidal alternating electric field if the frequency of the field exactly matches the cyclotron frequency.

Explain how this concept may be utilized to design a mass spectrometer?

8.4 Time-of -flight mass spectrometer

If the ions of mass m and charge e, which have been accelerated through a potential difference of V in the ion source, are allowed to drift in the field-free region, their velocity v will be governed by eqn (8.1); i.e.

$$v = \left(\frac{2eV}{m} \right)^{½}$$

Thus, the time required by the ions to traverse a distance, a, in the field-free region would be given by

$$t = d \left(\frac{m}{2eV} \right)^{½} \tag{8.5}$$

It is evident from eqn (8.5) that, for a beam of ions possessing the same kinetic energy, the time of flight, t, is directly proportional to the square root of m/e ratio.

Suggest the design of a mass spectrometer based on this principle. Mention the possible merits and demerits of the system.

ANALYTICAL ELECTRON MICROSCOPES

9.1 INTRODUCTION

Analytical electron microscopy is a rapidly developing new area within electron microscopy in general. It is important to the researchers and technologists who are concenred with the development and characterization of new materials. In order to determine the composition and structure of analytical species, it becomes necessary to collect several signals emitted by the specimen when it is irradiated by an electron beam. The instrument that has built-in capability to collect and analyze these signals is called an analytical electron microsope (abbrevaiated as AEM).

9.2 DESIGN CRITERIA

A number of signals as shown in Fig.9.1, are generated at each point when a fine electron beam strikes the specimen.

The mechanism of emission of the signals and the way, they may be used for analytical purposes are discussed, in brief, as follows.

Secondary electrons

As the electron beam enters the specimen, the electrons collide with the atoms in the sample and knock out their orbital electrons. Thus secondary electrons are generated throughout the capture volume, which is effectively confined to the surface layer of the specimen. The secondary electrons have very low energy (of the order of 50 eV or less) and hence they can travel only a short distance. Thus they come primarily from the region immediately below the impact point of beam and carry information about that region.

The surface topography may be employed for obtaining the contrast in the image pattern. A minor change in the surface angle relative to the incident beam causes an appreciable change in the number of secondary electrons that can escape; because the capture volume near the surface becomes more. Similarly, a sharp edge on the specimen exposes more area and thus emits a large number of secondary electrons. Thus, as the electron beam scans the specimen, the brightness of the cathode ray display tube changes gradually as the specimen slope varies and becomes very bright at the edge. The secondary electron images of the most surfaces can be easily understood by the operator, and hence very little interpretation is required for qualitative work.

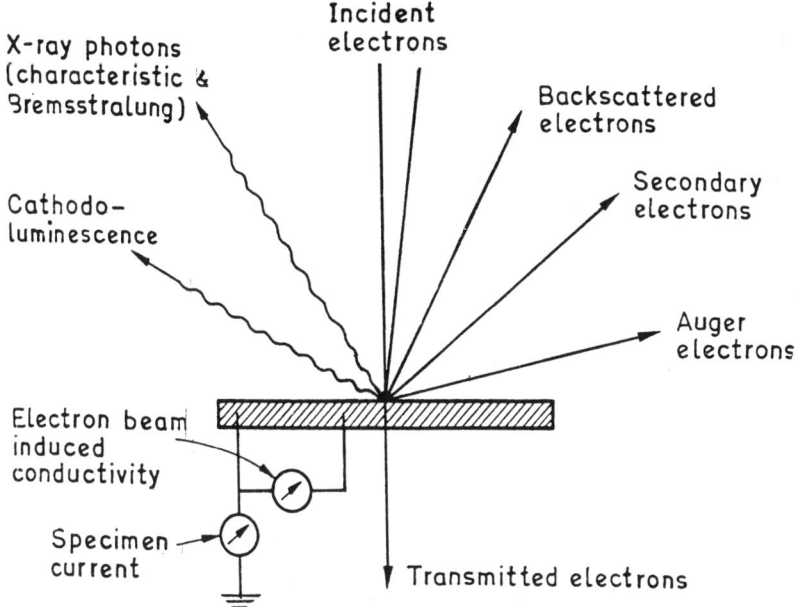

Fig.9.1 :   Various signals generated by the electron beam upon striking the specimen.

Backscattered electrons

If a highly energetic electron beam (of the order of 50 KV strikes the sample, the incident electrons may undergo a kind of Rutherford scattering from the sample atoms and re-emerge from the surface. The backscattered electrons still have almost the same energy and can be detected by Schottkey barrier photocell. The resulting image is, in some ways, analogous to the secondary electron image, but there are some important differences.

In this case, the incident electrons have greater energy and hence they can penetratae to a greater depth. The backscattered electrons, therefore, contain less information about the surface and more about the bulk material.. The backscattered electron signal is directly proportional to the average atomic number Z of the specimen. This fact can be used for elemental analysis in some cases.

If the sample is crystalline and the angle between incident beam and the lattice planes in the sample, permits Bragg diffraction to take place, the incident electrons penetrate more deeply into the specimen and hence very few electrons are backscattered to the detector. Thus if the beam scans over a perfect single crystal, the image of backscattered electrons is not uniform but it consists of dark diffraction lines running across it. These lines can be used to index the crystal structure and the orientation. These patterns are called pseudo Kikuchi patterns.

X-Rays emission

The high energy electrons upon striking the specimen may knock off electrons from the inner shells (i.e. K or L shell) of the atoms present. The vacancy of the electron in the inner shell is generally filled by a transition of electron from the higher shell. Such transitions give rise to emission of X-ray photons which are characteristic of the elements present in the sample. In addition, the decceleration of electrons also cause the emission of cont005 X-rays or Bremsstralung radiation. Thus an X-ray spectrum is obtained which consists of several lines superposed on a continuous background.

The X-ray signal can be utilized in many ways. Entire X-ray signal can be detected and displayed to give some information about the specimen, but the use of total X-ray flux to produce an image has been very rare. Instead, measurement of X-ray wavelength or energy to obtain analytical data is very common. The diffraction technique is generally not employed for wavelength measurement. A relatively recently developed technique, called energy dispersive X-ray analysis (EDXA) is often preferred. This is discussed in tutorial 9.5.

Auger electrons

An energetic electron upon striking the atom produces an ionization by knocking out a bound electron from the inner shell (to become a secondary electron), the decay produces an X-ray photon. This photon is reabsorbed by the atom to produce a photo-electron from another shell. Such photo electrons are called Auger electrons. There is high probability of emission of these electrons from very light elements and hence these may be used to determine the lighter elements.

Transmitted electrons

An energetic electron beam (accelerating voltage of the order of 50 KV) can easily pass through a specimen whose thickness is of the order of few μm. The transmitted electrons can be detected by a suitable detector (e.g. a scintillator-photomultiplier tube combination). The image so obtained is good enough to identify major features of the specimen.

A technique similar to a conventional transmission electron microscopoy (TEM) can also be employed to image the static patterns of interest.

By employing an electron spectrometer under the specimen to separate the electrons that have lost energy from those which have not and then electronically ratioing the inelastically to elastically scattered electrons it is possible to obtain an image signal that is dependent only on the atomic number. This technique is sensitive enough to detect the presence of heavy atoms on a substrate. The technique is called scanning transmission electron microscopy (STEM). It is possible to examine specimen with STEM several times thickner than that can be imaged by a conventional TEM of the same accelerating voltage.

Cathodoluminescence

The incident electron beam can excite luminescence in some materials, e.g., minerals, some organic compounds etc. and they may emit visible or nearly visible radiation. The emitted radiation can be easily detected by PMT. The intensity of this radiation is however insufficient for using a dispersion device to measure the wavelength. Nevertheless, the technique is very sensitive to changes in composition at the ppm level and even below.

Electron-beam induced conductivity

The deceleration of an energetic electron in a semiconductor may create an electron-hole pair. These free charge carriers can produce a current across an intrinsic junction. Thus it is possible to locate the junctions by scanning the electron beam over the sample and imaging the current flowing in the device. The bright lines on the image will identify the junctions.

Combined Techniques

As we have seen, various signals are emitted by the specimen when it is irradiated by an electron beam, and they carry information about the composition and structure of the specimen. The instrument that can collect and analyse these signals, is called an analytical electron microscope (AEM).

What should be the configuration of an AEM?

The first component that is required is a source of electrons. These electrons should have sufficient energy so that upon striking specimen, they may produce the desired signals. It means that some means of accelerating the electrons must be provided in the module producing electrons. Normally, the electron beam diameter is large and hence a module for demagnification of the spot size will be an integral part of the instrument. Further, a mechanism, for scanning the beam over the specimen surface, will be required. The electron beam after interaction with the sample will generate several signals. In order to collect these signals, relevent detectors will also be required. Finally, the signal processing and display devices will be needed for presenting the analytical information in the desired form. Accordingly, the modules in a generalized configuration of an AEM may be arranged as shown in Fig.9.2. The principal components, shown here, are discussed in brief, in the following section.

9.3 PRINCIPAL COMPONENTS OF AN AEM

The source of electrons is either a heated tungsten or lanthanum hexa boride emitter. The minimum size of the electron probe is of the order of 2 nm (20 A°). In advanced versions of AEM, cold field emission sources are employed. These sources are much brighter and the probe size is also smaller (typically about 0.5 nm at the specimen). The brightness of electron gun and the probe size determine the number of electromagnetic lenses needed to achieve a suitable demagnification of the electron beam. However, there is no thoretically optimum number of lenses.

For scanning, two (or more) deflecting coils are used so that the beam can strike at a single point on the specimen and rock. The deflecting coils can also vary the angle of incidence for diffraction studies. Either continuous (or ramp) scan generators which are generally more reliable or step scan generators (or ladder circuits) that facilitate computer interfacing may be employed. Most instruments, however, offer a TV rate scaning capability for real time examination of specimen. The common display devices are long persistence (slowly scanned) cathode ray tubes or conventional TV monitors. Besides, provisions to accomodate large specimen, and to move them in at least two directions and tilt, heating and cooling facility, rapid specimen exchange with airlock facility and externally controlled manipulators are also part of the equipment.

As the scattering and absorption of electrons in air must be avoided, a vacuum system is a must for all the electron beam instruments. The trend is towards the use of a cleaner vacuum system both to reduce the contamination of the specimen and also to permit the use of brighter guns.

For microanalysis, two spectrometers are necessarily fitted with AEM and they are located in different positions on that instrument. The energy dispersive X-ray analyzer (EDXA) is used to detect the presence of heavier

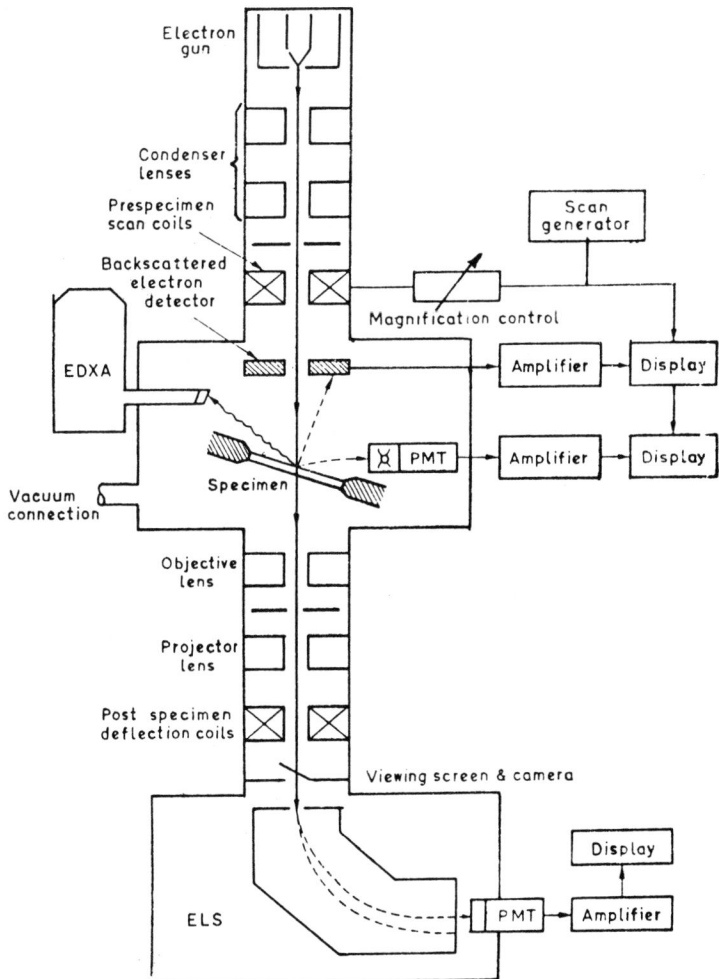

Fig.9.2 : Schematic diagram of an analytical electron microscope.

elements in compounds and alloys. The electron energy loss spectrometer (ELS) is employed for detecting lighter elements. The design of ELS is based on the electromagnetic or electrostatic deflection (or both) of the electron beam after transmission through the specimen. The magnetic sector (similar to that employed in a mass spectrometer) without retarding field is generally employed for dispersing the electrons of different energies. Typical design employs a soft iron pole pieces with a 90°bending angle and radius between 10 and 25 cm. Typically, a dispersion of 3μm per eV loss may be produced. The energy resolution can be controlled by the slit system of the spectrometer with the range of 1 to 30 eV. The dispersed electrons are moved across a fixed detector slit by ramping the magnetic field or by using a set of scanning coils. A combination of scintillator and PMT is employed for detection.

9.4 ANALYTICAL APPLICATIONS

The microstructures of high performance materials are quite complex and multiphase. So, in order to investigate such materials, a instrument is required which has both local composition and local structure determination capabilities

For example, elemental analysis may be very useful for distinguishing between certain refractory metal carbides and nitrides (and oxy-carbonitrides) that are isostructural and have minor differences in lattice parameters. EDXA and ELS attachments of an AEM can perform elemental microanalysis. The structural details can be obtained by diffraction imaging. The surface studies can be performed through secondary electrons and so on. Thus an AEM is a modern tool for microanalysis.

TUTORIAL-9

9.1 An analytical electron microscope (AEM), which can operate simultaneously in the following modes, is to be designed. Suggest the schematic arrangement of basic components for each mode of operation separately and that for all the modes operating simultaneously.

(a) *Emissive mode :* in which the secondary electron emission or the backscattered electron beam is used to obtain information about the nature of the specimen surface.

(b) *Conductive mode:* in which the electron beam induced currents (in the specimen circuit) are used to obtain information about the processes occuring under the surface.

(c) *Luminescent mode:* in which the recombination radiation induced in the specimen as a result of the primary beam bombardment is employed to obtain knowledge of specimens which have been damaged and which can not be employed in the electrical circuit.

9.2. How, by use of optical coupling, a low noise wide band width system can be designed for the detection of the secondary electron emission in an AEM?

Hint: The detector for the secondary electrons consists of a grid (generally called a Faraday cage) which is held at a high positive potential (e.g. 10 KV) in front of a scintillator or phosphor as shown in Fig.9.3. The secondary electrons emitted by the sample, are attracted and accelerated by the field of the Faraday cage; travel in curved trajectories and strike the scintillator. The resulting light is conducted through a light pipe to the photomultiplier tube. The signal after amplification can be dealt with by conventional electronics.

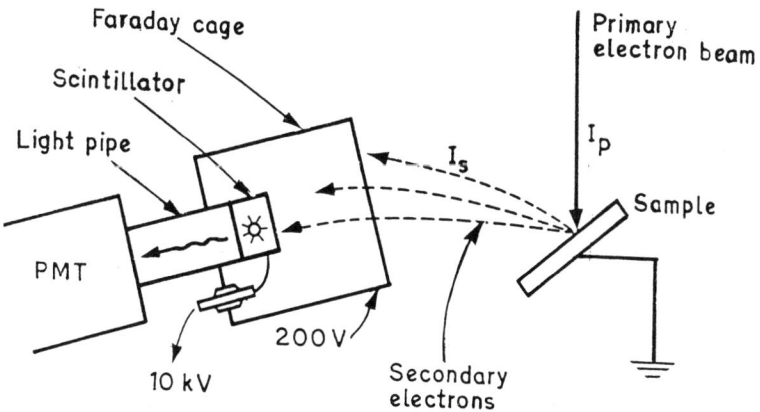

Fig.9.3 Schematic diagram of detector for secondary electrons

9.3. How is it possible to use a specimen containing a p-n junction to act as its own detector in an AEM?

Hint: It is possible to use a specimen containing a p-n junction to act as its own detector. Light resulting from the beam injected carriers is propagated internally and assumed to be reabsorbed in the junction layer due to changes in absorption coefficient and so gives rise to a photocurrent (see Fig.9.4).

9.4. How the leakage current to earth can be utilized to obtain information about the specimen in an AEM? Discuss the factors which may lead to the contrast in this mode of operation.

Hint: Leakage current to earth can be utilized to obtain information about the specimen as shown in Fig.9.5. Factors which affect both the reflection of primary electrons and the emission of secondary electrons can lead to contrast in this mode of operation.

9.5 Can X-ray analysis of a sample be performed employing SEM or STEM? What additional components would be necessary for such an analysis?

Hint: The module performing the X-ray analysis is shown in Fig.9.6. It employs a recently developed technique of energy dispersive x-ray

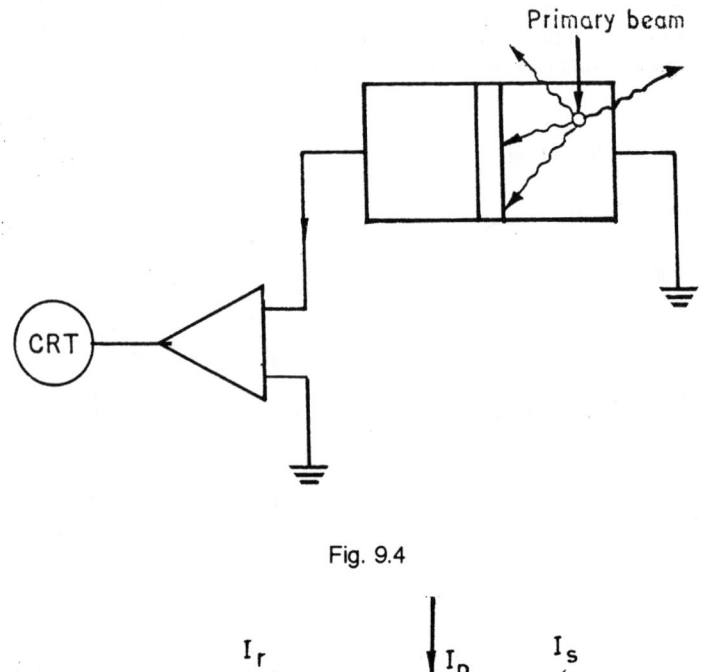

Fig. 9.4

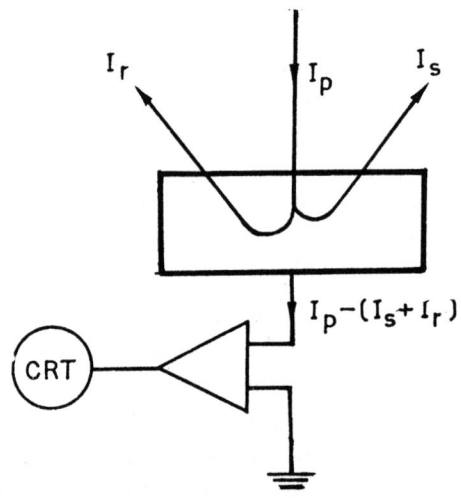

Fig.9.5

analysis (EDXA). It is so called because it measures the energy of the X-ray photon, rather than its wavelength. A pure wafer of silicon acts as solid state ionization chamber (detector). The incident X-ray photon creates a photoelectron and subsequent ionization of the silicon atom. As each ionization requires 3.8 eV, the number of silicon atoms ionized and the number of electrons freed depend linearly on the energy of X-ray photon. The total charge deposited in the silicon detector is collected by applying a bias voltage. The charge is integrated to give a small pulse. All these operations are linear and hence the pulse height is a measure of energy of the X-ray photon. These pulses are measured by a multichannel analyser and the spectrum displayed either on CRT or TV monitor.

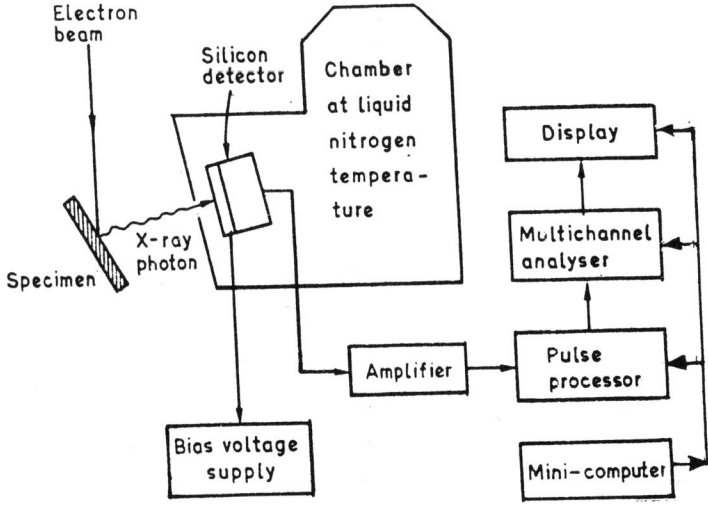

Fig.9.6: Schematic diagram of EDXA for AEM

10
ON-LINE ANALYSERS

10.1 INTRODUCTION

The development of new materials requires a very precise investigation of their chemical composition and their physical properties. Such a qualitative material characterization can be done only in an off-line labor atory employing one or more analysis instruments (which have been discussed in earlier chapters). However, the manufacture of materials may not require any qualitative exploration but a close control of the product quality is always essential. The instrumentation requirement of the production unit, therefore, is that of routine or quantitative on-line analysis and process control. On-line analysis involves the measurement of the properties of the product right at the point of production as opposed to removing the samples from the process stream from time to time and analysing them in a laboratory.

What should be the configuration of an on-line analyser? A complete on-line analysis instrument would consist of, at least, two parts; viz.

(i) the sampling system for obtaining a representative sample from the process stream and conduct it to the analyser, and

(ii) the analyser itself, which can perform the measurement and send the information at the desired time intervals.

For a variety of purposes, particularly, industrial safety, control of environment, control of chemical and metallurgical process and research, a large number of analysers have come into existence. These are based on various physical or physico-chemical methods. Some of the common on-line analysers have been discussed in this chapter. Before we take up these systems let us discuss, in brief, the sampling systems required for these analysers.

10.2 SAMPLING SYSTEMS

The object of the sampling system is to obtain a truely representative sample from the on-line process stream, process it to the required physical and chemical state without removing its essential ingradients and conduct it to the analysis instrument continuously or periodically. Thus the sampling system consists of a sampling tube which withdraws the sample from the pipe or the flue and the sampling line which carries the sample from the sampling tube to the analyser. Most of the analysers are designed for operation at or near atmospheric pressure and the process fluids are, generally, at different pressures and hence it is essential to provide

(i) an aspirator or pump which can draw the sample from the process fluid, if it is below the atmospheric pressure or

(ii) a let-down valve or capillary to reduce the pressure to a desired value, if the process gas or liquid is at a higher pressure.

In addition, a number of other components such as filters for dust removal, condensate traps, relief valves, cooling systems, drying chambers and so on, may also be required.

While designing a sampling system for on-line analysis, it must be kept in mind that the time interval between the occurrence of a change in the process stream and its detection at the analysing instrument be as short as possible.

10.2.1 Sampling systems for gases

The components of a sampling system for gases fall into two categories. (1) Initial components, which condition the gas (to be analysed) to the required state so that it can be passed through the sampling line. They may include a sampling tube or probe, coarse filter, cooler or condensate trap, pressure regulator or pump to reduce or increase pressure and so on. (2) Final sample system, which is generally mounted with the analyser and which may be required to control the flow of the gas through the analyser , to calibrate or select the stream (in case of multistream analysis). Thus the components of a final sample system may be a fine filter, flow control valve and flow meter, calibration valve or by-pass split and so on.

The materials for these components must be such that they withstand both mechanical and thermal shock; do not react with, absorb or cause catalytic reaction in nor be corroded by the sample gas. In general, non-corrosive gases are handled by systems employing copper, aluminium or plastic tubing with brass or nickel plated brass components and fittings. Corrosive gases are handled in stainless steel with teflon. Glass or plastic is commonly used in catch pots and flow meters.

Sample Probes

Sample probes are employed for sampling gases from a closed vessel or a wide pipe. Their main function is to position the sampling point in the main body in order to obtain a representative sample. The probes are not required for sampling homogeneous gases in narrow pipes or for atmospheric air. The secondary use of the probe is to give the sampled gas a preliminary cleaning. Fig.10.1 shows a straight forward open probe for relatively clean gases. This is a simple tube fitted with a flange for mounting and a plug for rodding (i.e. cleaning with a rod). An elbow at the end of this tube prevents the large particles being swept by the gas flow, but it makes rodding difficult.

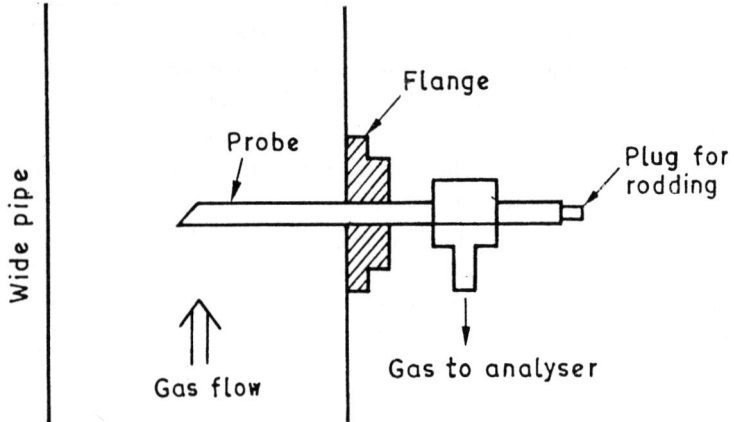

Fig.10.1 Open probe for sampling gases from a wide pipe

Upto a temperature of 800-900° C, stainless steel probes without cooling and above this temperature upto about 1600°C, water -cooled probes are generally employed. For still higher temperature, ceramic or silica probes may be used.

For dirty gases, water washed probes are employed. A spray of water is spread across the opening of the tube which washes the dirt entering it. This method cuts down corrosion by washing out sulphur dioxide and sulphur trioxide at source. It is generally employed for analysis of oxygen but is unsuitable for the gases soluble in water.

Filters

The particulate matter can adversely affect the functioning of the analyser and hence it must be removed from the sampled gas. Every analyser should have a high efficiency filteration equipment. The filters having 99.97% efficiency at 0.3μ(mesh) are easily available. A typical coarse filter is shown in Fig.10.2. A sintered disc is acting as a filter element and is located at a point where the bypass and the sample flow split.

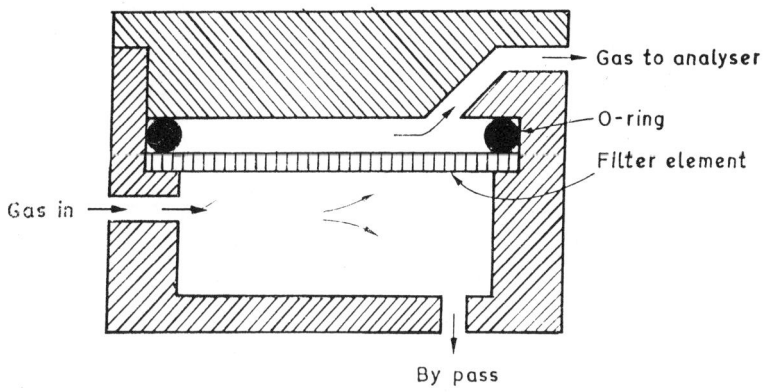

Fig. 10.2 Schematic diagram of a coarse bypass filter

Some filters employ a cyclone action in which the centrifugal force keeps the larger particles and water droplets away from the annular filter element. These are used for heavy dirt conditions. For very fine dust, electrostatic filters may be used but they are expensive as well as relatively bulky.

Cooling and condensate removal system

If the sampled gas is at a higher temperature, it should be cooled before sending it to the analyser. For dry gases cooling alone is not a problem but if the gas contains condensible vapours, the cooler should be accompained by the condensate removal system, otherwise the condensate may cause several problems e.g., blockage of sample lines, damage to analyser cell, disturbing the output and so on. Some of the systems for cooling and condensate removal are discussed as follows.

Air cooling shown in Fig.10.3(a) is easy to install but is least efficient. Water-cooled cooler shown in Fig.10.3(b) is most common but may not always be practical or economical. For higher efficiency, a vortex cooler using compressed air or an electric cooler may be employed.

The condensed water or other condensate must be drained away. If the content of the condensate and its flow rate is low it can be drained manually by operating a valve at reasonable intervals. If the sampled gas has been cooled in the sample line, the condensate (moisture) can be removed by employing a lute or catch pot shown in Fig.10.4.

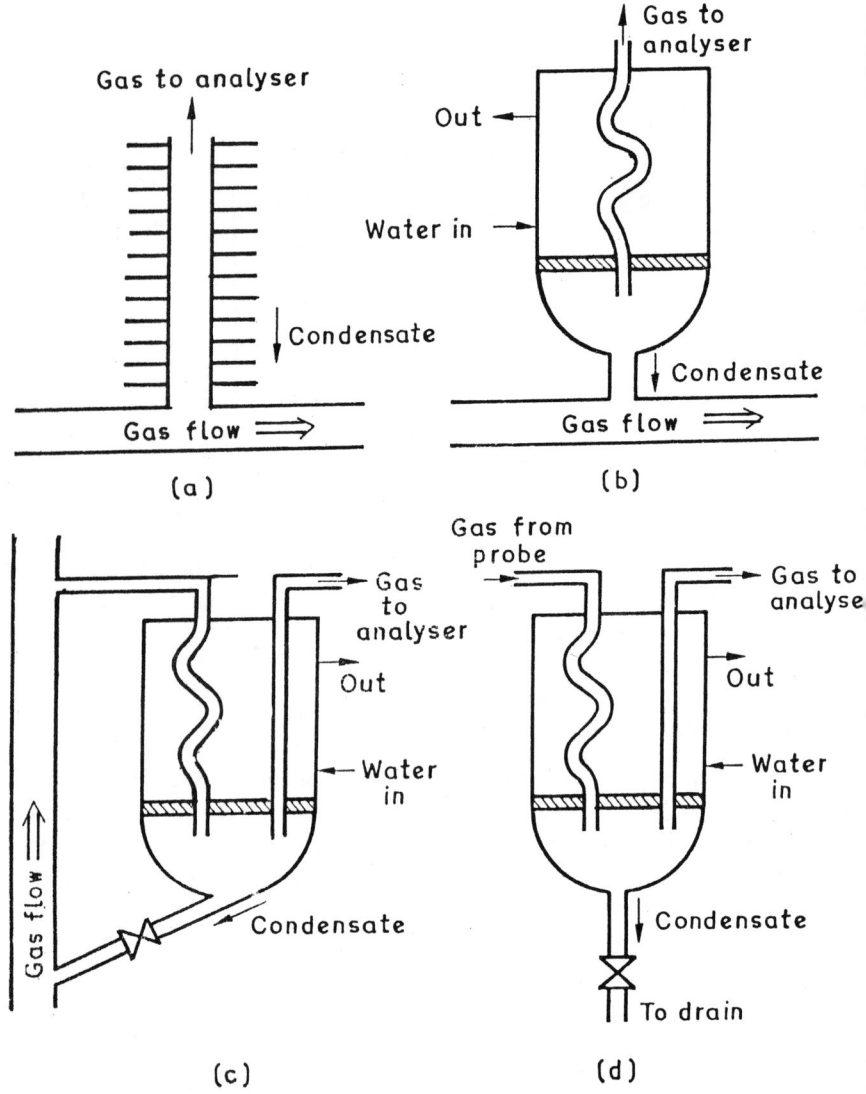

Fig.10.3 (a) Air cooler,
         (b) Water-cooled cooler on horizontal narrow pipe
         (c) water-cooled cooler on vertical narrow pipe and
         (d) water-cooled cooler for a gas from wide pipe.

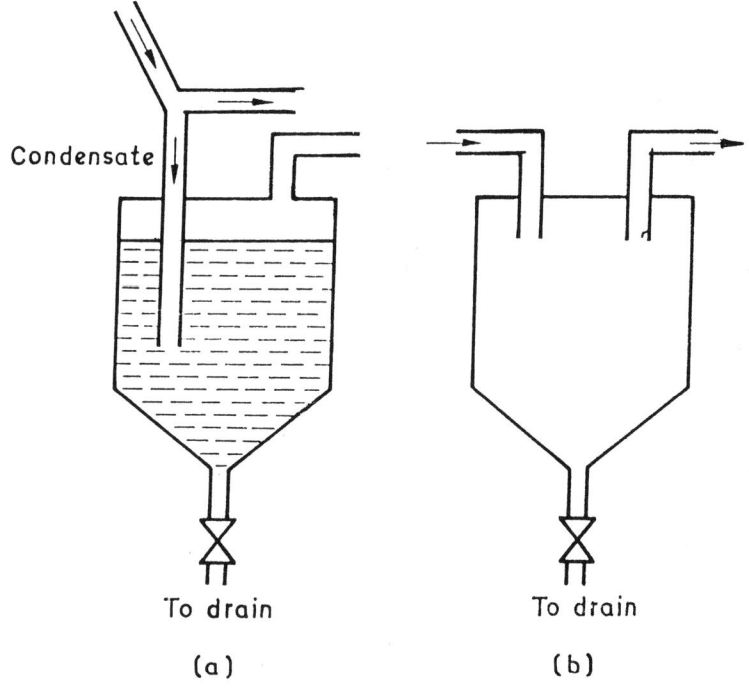

Condensate

To drain

(a)

To drain

(b)

Fig.10.4 (a) Lute and
(b) Catch pot

Pressure adjustment and control system

If the sampled gas is at a higher pressure at the sampling point, its pressure must be reduced to suit the final system; and if the gas is at a lower or atmospheric pressure, it must be pressurized so that it can be pushed through the sample line and the analyser. Thus the pressure regulators, electrically driven pumps, and ejectors or aspirators operated by water, steam or air are essential components of a sampling system.

Final sample system

The final sampling system completes the conditioning of the sample gas and controls its flow through the analyser. For single analyser, the system would normally be as shown schematically in Fig.10.5. The gas from the initial sample system enters the final sample system through a pump or pressure regulator and the flow is controlled by the needle valve. The bypass is almost aways employed. A fine filter immediately before the analyser is invariably required to protect the analyser from fine dust. The flow meter either before the analyser or after it can be advantegous but is not essential.

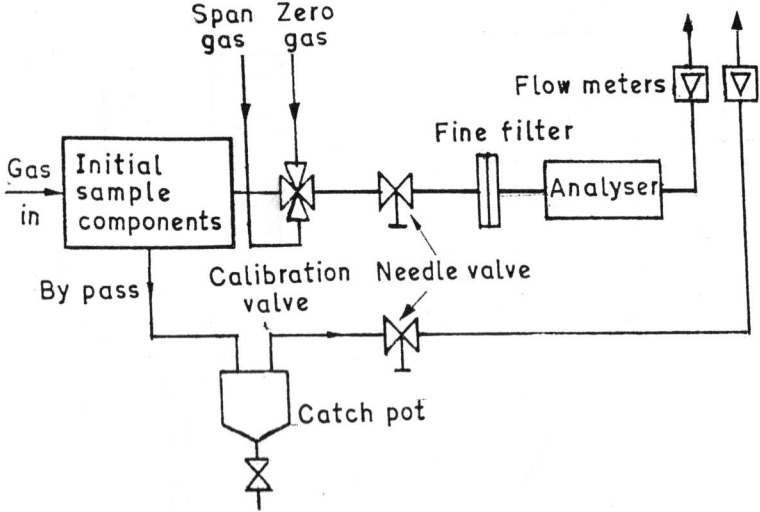

Fig.10.5 Schematic diagram of final sample system

10.2.2 Sampling systems for liquids

The methods of sampling liquid process streams are almost similar to that of gaseous streams. A typical system is shown in Fig.10.6. In order to obtain a representative sample, the process fluid should be thoroughly mixed. For this purpose an in-line mixer is generally installed. This consists of from four to six curved vanes arranged such that they produce rotation of the liquid stream in opposite directions. In some cases, the pipe section is enlarged and an impeller driven by the motor is fitted. Its axis of rotation is at right angles to the direction of flow. Generally the sample is removed from the centre of the pipe as shown in Fig.10.6. The sample lines to the analyser should be as short as possible. If required, a fast loop may be designed with a pump to circulate the process fluid past the analyser via a bypass filter. After passage through the analyser, the sample may be collected in a suitable drain tank and pumped back into the return line at suitable intervals.

The sample conditioning equipment, in this case is nearly the same as that used for gases, but the difficulty, generally, arises in obtaining a truely representative sample. The tendency, therefore, is where possible to adopt techniques of analysis which do not require the sample to be withdrawn from the process line. For example,. an immersion or dip type assembly that is used in pH or electrical conductivity measurements, may be used.

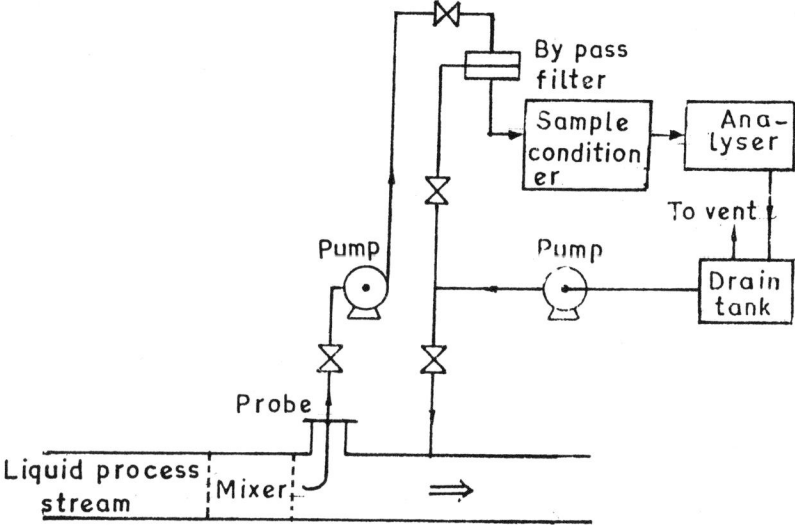

Fig.10.6 Schematic diagram of sampling system for liquids

10.3 NON-DISPERSIVE OPTICAL ANALYSERS

Of the several optical methods available, the absorption technique has been found to be most suitable and is widely used in on-line analysis of gases and liquids. Non-dispersive type of instruments are more common than the dispersive instruments employing prisms or gratings. All the common gases except O_2, N_2, H_2, Cl_2 and the inert gases and nearly all the organic vapours can be analysed by measuring their absorption in the infra-red region. Although UV/VIS analysers are less common, they can be employed for analysing liquids as well as gaseous samples.

10.3.1 Non-dispersive Infrared (NDIR) analysers

The NDIR analysers are based on a simple principle. When the radiation from a source emitting IR is passed through a sample cell, it may be absorbed if the infrared absorbing component is present in the sample gas. Thus the intensity of the radiation coming out from the sample cell can be calibrated interms of or used to monitor the concentration of the absorbing component in the gas.

A typical double beam instrument based on the above principle is shown in Fig.10.7.

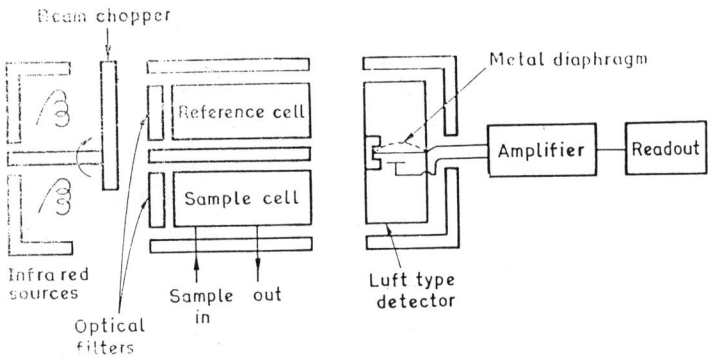

Fig.10.7 Schematic diagram of a double beam NDIR analyser

The radiation from the ir sources (mainly heated filaments) is chopped by a mechanical beam chopper. Thus the radiation passes alternatively through the sample and reference cells to the detector. Optical filters are not essential but they are important in narrowing down the spectral range and thus improving the selectivity. Identical filters are employed in the two cells . The detector consists of two chambers with a metal diaphragm between them. The chambers contain the gas which absorbs at the same wavelength as the sample gas. Generally, it is the same gas. There is arranged a metal plate near the diaphragm to form a capacitor like arrangement. So long as the fraction of the component gas that is being monitored in the sample remains constant, the intensity of the radiation reaching the two chambers of the detector will be constant. Any variation in the concentration of the observed component will change the intensity of ir radiation reaching the sample chamber of the detector. This will create unbalance in the pressure of the gas in two chambers thus pushing the metal diaphragm to one side. As the distance between the diaphragm and the plate changes, the capacitance also changes. Thus an ac signal is obtained whose frequency is related to the chopper speed and the amplitude is proportional to the concentration of the analytical species in the sample gas. The reference cell generally contains nitrogen or a mixture of background gases in nitrogen that are arranged to give the same total absorption as the sample stream so that the optical system is balanced. The capacitance variation of low frequency can be picked up directly as a voltage variation or it may be used to amplitude or frequency modulate a RF carrier. Then it is amplified and converted to a d.c. signal for presenting on a meter or a recorder.

Compact analysers using solid state detectors have recently become available. As the size of the detector is small, the source radiation has to be focussed onto it employing either mirror or lens optics as shown in Fig.10.8. Instead of two, a single optical filter immediately before the detector is sufficient.

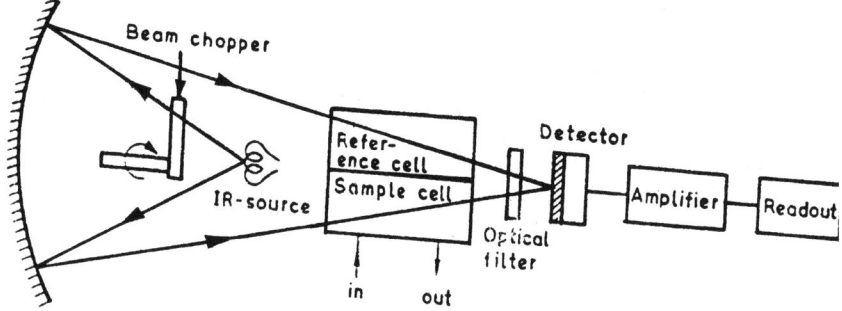

Fig.10.8 Schematic diagram of a solid state detector NDIR analyser

10.3.2 UV/VIS Analysers

As compared to NDIR analysers, very few UV/VIS analysers are in use. However, they are versatile and can handle liquid as well as gaseous samples. Other properties such as film thickness, colour, turbidity, etc. along with the chemical composition can also be measured or monitored. Here also a reference signal is necessary but instead of a double cell, a single cell design is preferred. After the light beam passes the sample cell, it is split into two beams which go through the sample and reference filters as shown in Fig .10.9. The sample filter selects the wavelength at which the component of interest absorbs and the reference filter selects that wavelength which is not absorbed. The detectors are often phototubes, that are used with logarithmic amplifiers to give a wide linear range.

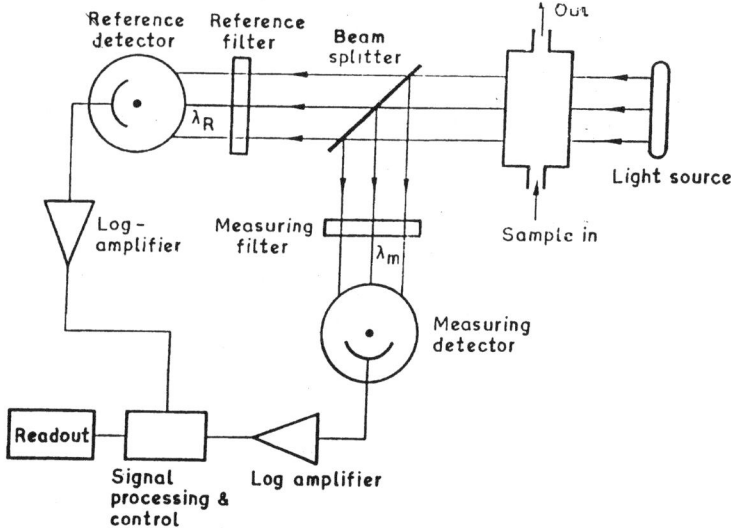

Fig.10.9 : Schematic diagram of dual wavelength UV/VIS analyser for liquids & gases (Du pont photometric analyser is based on this principle)

10.3.3 Process refractometers

The refractive index of a given medium varies with density, temperature, wavelength of the incident radiation, pressure, and so on. Therefore if these variables are kept constant, the refractive index becomes a characteristic constant of the medium.

The instruments that are designed for continuous and automatic recording of refractive indices are called process refractrometers. A typical refractometer is shown in Fig.10.10. This instrument is employing a servomechanism to maintain a constant position of the image on the twin detectors. The light from the source is collimated by the lens and chopped by a rotating sector (not shown), then passes through a double prism cell. The refracted beam is focussed on the mirror which images it on the twin detectors. So long as the composition of the fluid flowing through the sample and reference cell is same, equal light falls on the two detectors. This is equilibrium position. Any change in the sample process stream causes imbalance which is removed by the servomotor geared to the mirror. The rotation of the mirror is made proportional to the refractive index of the sample stream and is correlated with known standards.

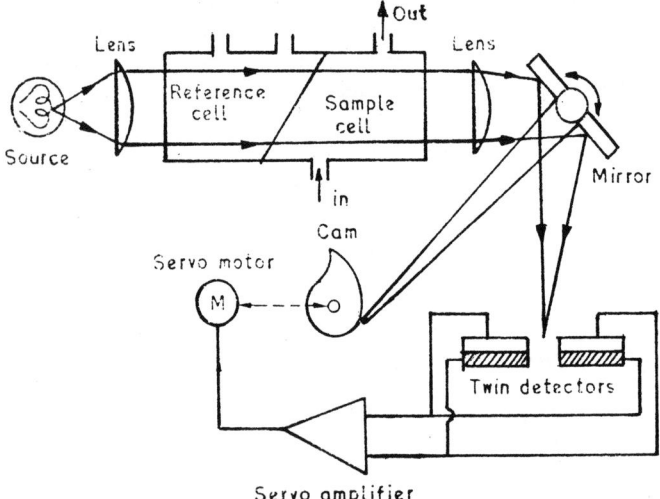

Fig.10.10 Schematic diagram of a process refractometer

Such refractometers have been used in sugar concentration analysis, determining the consistency of ketchup, controlling the blending of styrene and butadiene streams in the manufacture of GR-S rubber, brewing industries and so on.

10.4. THERMAL CONDUCTIVITY GAS ANALYSERS

The simplest method of continuous gas analysis is to install an electrically heated filament in the gaseous stream and monitor its resistance. If the flow rate of the gas through the cell containing filament and the current through the filament are maintained constant, the loss of heat from the filament and hence its temperature will depend on the transfer of heat to the gas. This will depend, among other things, on the thermal conductivity of the gas and therefore, on its composition. The change in temperature of the filament causes the change in its resistance. Thus a Wheatstone's bridge circuit can be used normally with a comparison filament, to monitor the changes in the resistance of the measuring filament. The thermal conductivity (TC) analysers are based on this principle. A TC cell is also called a katharometer.

In general, for a thin filament inside a TC cell, the rate of heat loss by the hot filament may be related to the thermal conductivity (k) of the gas flowing through the cell, by the expression:

$$\frac{I^2 R}{J} = kA \frac{dT}{dx} \tag{10.1}$$

where I is the current flowing through the filament of resistance R, J is the mechanical equivalent of heat, A is the cross-sectional area of the surface normal to the heat flow and dT/dx is the temperature gradient formed with distance x from the filament.

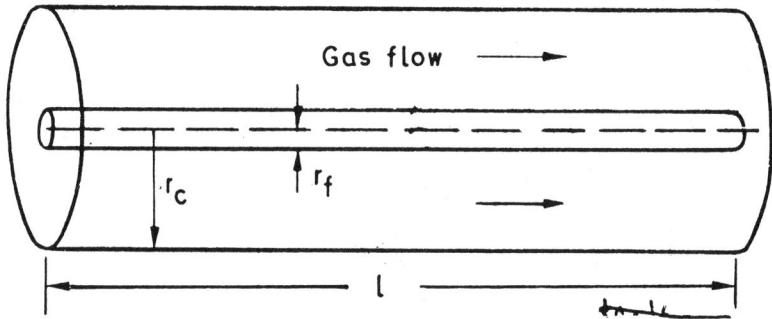

Fig.10.11 : Straight wire filament inside a cylindrical cell

For a straight wire filament of length l, radius of cross-section r_f and temperature T_f, that is placed along the axis of the cylindrical cell of radius r_c and cell wall temperature T_c (as shown in Fig.10.11) equation (10.1) modifies to

$$\frac{I^2 R}{J} = \frac{2\pi kl (T_f - T_c)}{\ln (r_c/r_f)} \tag{10.2}$$

or $$\dfrac{I^2 R}{J} = kG. \Delta T \qquad\qquad (10.2a)$$

where $G = \dfrac{2\pi l}{\ln(r_c/r_f)}$ is known as the cell factor

and $\Delta T = T_f - T_c$

Thus, if T_c is kept constant, the rate of heat loss is directly proportional to k and hence to the gas composition.

The design of a TC cell is important in deciding the sensitvity and the response of the analyser. Some common designs are illustrated in Fig.10.12.

A flow-through cell (Fig.10.12a) has a fast response but is highly sensitive to variation in the flow rate of the gas. A diffusion cell (Fig.10.12b) has a very slow response but is least sensitive to flow variation. Convection-exchange type of cell shows a better response as compared to convection - diffusion type of cell. Infact, commercial designs are a compromise between flow-through cell and the diffusion type of cell.

For absolute measurement of thermal conductivity a very careful control of all the relevent parameters e.g., current flowing through the filament or the voltage drop across the filament, temperature of the cell wall, flow rate of the gas, etc, is essential. In production units, however, absolute measurements are not necessary. A differential method is sufficient for on-line analysis. Herein, the resistance of the filament of the sample TC cell is compared with that of the reference cell containing a reference gas, through a Wheatstone bridge circuit. The reference gas must be stable and preferably should have the thermal conductivity close to that for the sample gas.

Two circuits, viz :

(i) employing one measuring cell and one reference cell; and

(ii) employing two measuring cells and two reference cells;

are shown in Fig.10.13(a&b). If a constant voltage V is applied to the bridge of Fig.10.13(a), and if all the bridge resistances have the same value, R, the unbalance signal, e, resulting from the change in resistance R of the sample filament is given by*

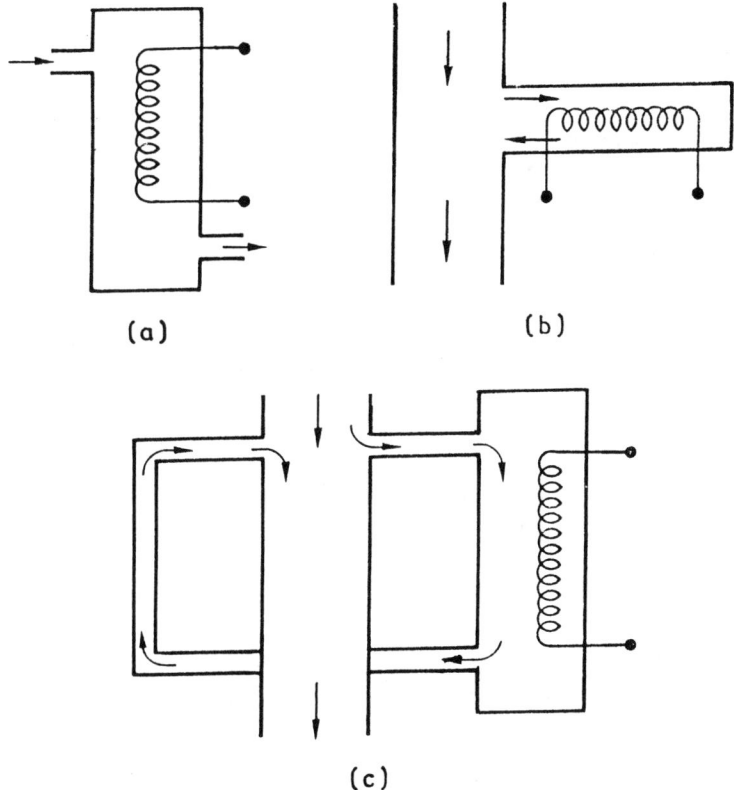

Fig.10.12 : Typical TC cells
  (a) flow-through cell,
  (b) diffusion cell, and
  (c) convection-exchange cell.

$$e = \frac{V \, \Delta R}{4 \, (R + \Delta R/2)} \tag{10.3}$$

For the bridge of Fig. 10.13(b), this expression modifies to

$$e = \frac{V \, \Delta R}{2(R + \Delta R/2)} \tag{10.4}$$

The response from the bridge can be optimized if all the four bridge resistances are nearly equal. As the thermal conductivity of the gas is temperature dependent, the control of the cell wall temperature to within ±

---

*A.E. Lawson, Jr and J.M. Miller: J. Gas chromatog: 4, 273 (1966).

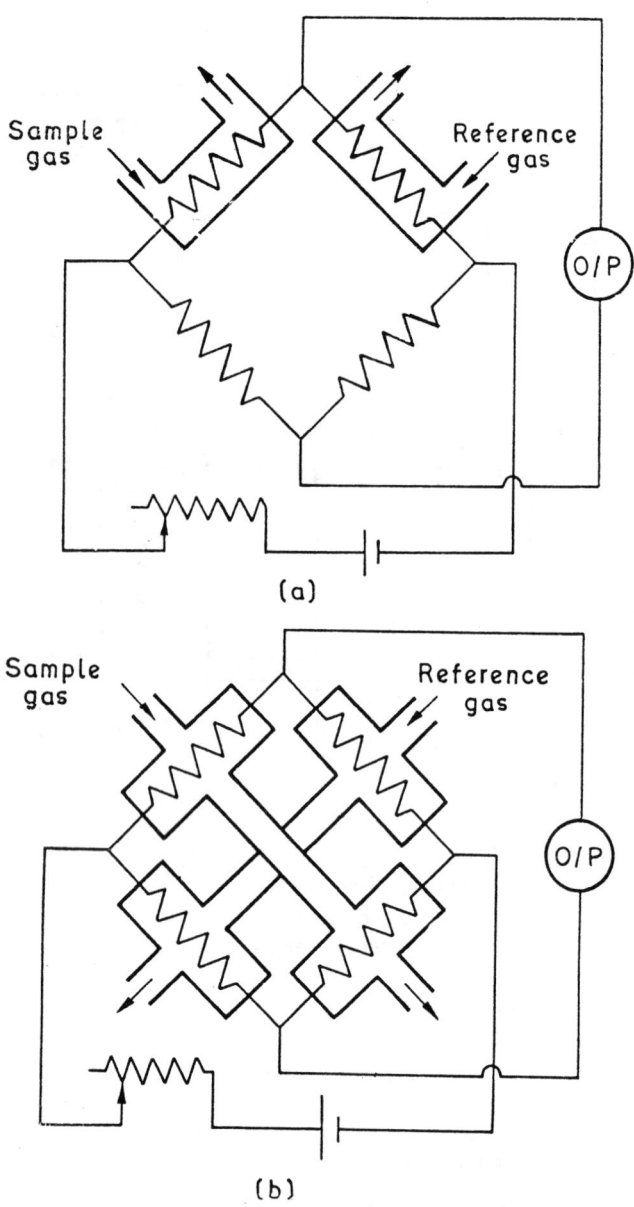

Fig.10.13 :   Typical circuits for TC analysers
   (a)   employing one sample cell and one reference cell and
   (b)   employing two sample cells and two reference cells.

0.1° C is necessary. Another point to be mentioned here is that while deriving equation (10.2) it has been assumed that the heat loss by the filament is primarily because of conduction. In practice, however, the heat loss, though small, may also occur through convection and radiation. Furthermore, equation (10.2) assumes that the filament is a thin solid cylindrical wire. In commercial designs, though, coiled-coil filaments are quite common.

A number of commercial TC analysers have been developed. Apart from measuring the single gas components, they can be used for analysing binary mixtures or psuedo-binary mixtures (e.g. CO_2 in air or flue gas, hydrogen in air or blast furnace top gas, water or SO_2 mixed with dry air and so on). Thermal conductivity detectors are widely used in gas chromatographs.

10.5 PARAMAGNETIC OXYGEN ANALYSERS

It has been found that the paramagnetic gases in an inhomogenous magnetic field tend to move from weaker to the stronger part of the magnetic field. The volume susceptibility (which is the ratio of the intensity of magnetization induced in the substance by the field and the field strength) of such gases is inversely proportional to the square of the absolute temperature. These properties have been utilized in designing the analysers for oxygen and other paramagnetic gases, e.g., nitric oxide, nitrogen dioxide, chlorine dioxide, etc.

10.5.1 Thermomagnetic oxygen analysers

Most of the oxygen analysers in use are based on what is known as the thermomagnetic method. Herein, the oxygen or the paramagnetic gas to be analysed, is allowed to pass through a cell containing a heated element in a magnetic field. During its passage through the hot zone, the gas is heated and it becomes less paramagnetic. Therefore, it is displaced by more strongly attracted cooler gas. Consequently, a continuous flow of gas results in the presence of a magnetic field. This is known as "magnetic wind". The flow rate is dependent mainly on the temperature of the hot zone, the strength of the magnetic field and the concentration of the paramagnetic gas, e.g., O_2.

A typical oxygen analyser based on this method is shown schematically in Fig.10.14. It consists of an annular cell around both sides of which the gas flows. Along the horizontal diameter there is a glass tube connecting the two sides, with two heating elements wound round it, these elements also serve as temperature sensors in the two arms of the Wheatstone's bridge . One of the elements is surrounded by a strong magnetic field , as shown. When the cross-tube is exactly horizontal and the paramagnetic gas, e.g. oxygen is absent in the gas, there is no flow across the tube. But, if the oxygen is present, this being a paramagnetic gas, is attracted from the left hand passage into the magnetic field. As it enters the tube, it is heated and becomes less magnetic and is, therefore, displaced by more magnetic cooler gas. Thus a continuous flow, a magnetic wind, from left to right results. This flow causes the transfer of heat from the left hand winding to the right hand one which consequently changes the temperatures and hence the resistance of windings. There is, therefore, an unbalance in the Wheatstone's bridge, in which they are connected. The output from the bridge is generally calibrated proportional to the oxygen concentration .

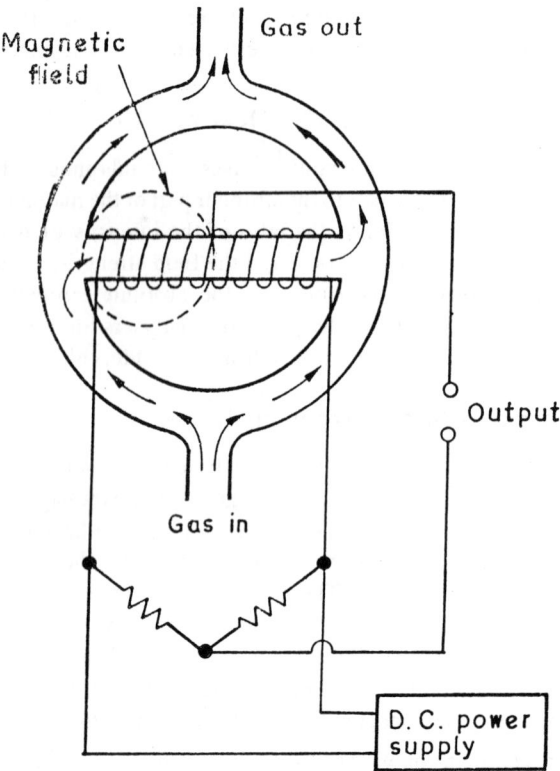

Fig. 10.14 : Schematic diagram of a magnetic wind oxygen analyser with a ring chamber.

This type of instrument has some limitations.

   (i) It is 'tilt' sensitive; and hence, the cross tube must remain horizontal during its operation in order to avoid the gravitational flow of the gas.

   (ii) The calibration of the instrument is valid for only one gas mixture.

(iii) The output signal becomes non-linear particularly at high oxygen concentrations.

The changes in sensitivity due to tilting are relatively less in the filament type of analyser shown in Fig.10.15. Herein, four heating elements form the four arms of the Wheatstone bridge, two of which are placed in the strong magnetic field as shown in Fig. 10.15(a). If the sample gas does not contain oxygen, the thermal convection generated by each filament is equal. However, if the gas contains oxygen, the convection in the magnetic arms increases because of the magnetic wind effect. Thus the filaments in the magnetic arms become cooler as compared to other arms thus unbalancing the bridge. The output can be calibrated in terms of the percentage of oxygen. In this device, a prior calibration using a reference gas is essential. A better design is possible with a reference gas flowing continuously through two arms as shown in Fig.10.15 (b). The reference gas is normally air (i.e. 21% O_2) or pure oxygen (100% O_2).

10.5.2 Magnetodynamic analysers

The thermomagnetic analysers are very sensitive to the changes in the gas composition, particularly those changes which involve large variation in the thermal conductivity. In order to overcome this limitation, another class of oxygen analysers has been developed. These analysers are based on the following principle.

When a diamagnetic test body is placed in non-uniform magnetic field of gradient $\dfrac{\partial H}{\partial s}$, a repelling force, F, (in the direction from the stronger to the weaker part of the field) acts on it and is given mathematically by*

$$F = 4\pi H \; \frac{\partial H}{\partial s} \; (k_1 - k_2)V \qquad (10.5)$$

where k_1 and k_2 are the magnetic susceptibilities of the test body and the gas surrounding the test body respectively; s denotes the distance, H is the magnetic field and V is the volume of the test body.

In a Pauling Cell, that is based on the above principle, two diamagnetic spheres (made of glass and filled with nitrogen) are mounted at the ends of a thin bar to form a dumb-bell. This dumb-bell is suspended horizontally with the help of a torsional fiber inside a cell which is filled with the sample gas. Inside the cell, there are magnets which produce a strong non-uniform magnetic field. The configuration is so arranged that the two diamagnetic spheres are repelled in opposite directions. Thus a torque acts on a dumb-bell

---

* D.J. Huskins: Quality measuring instruments in on-line process analysis (Chapter 6, p.238). Ellis Horwood Ltd, Chichester (1982).

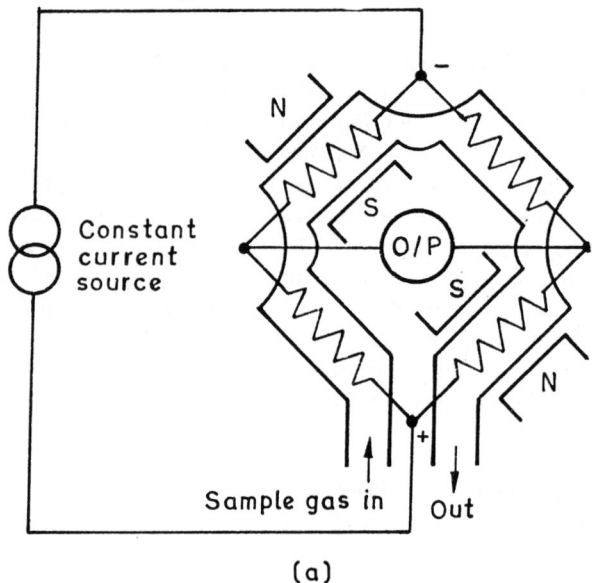

(a)

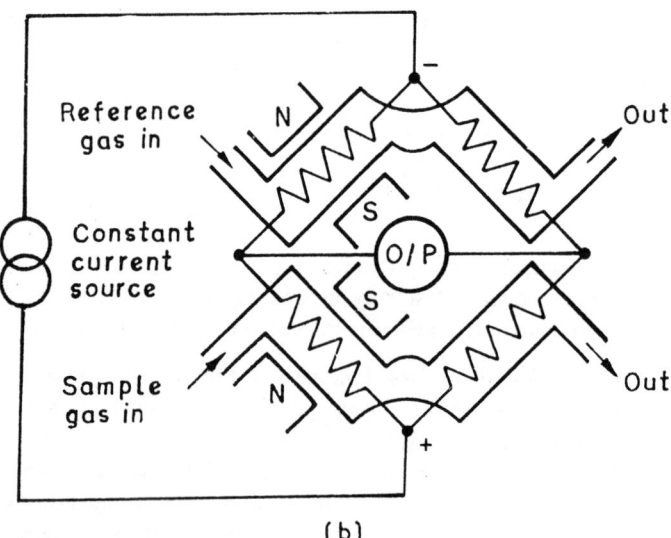

(b)

Fig. 10.15 : Schematic diagram of filament type magnetic wind analysers.
(Siemens oxygen analysers are based on this principle).

which rotates it and brings it to a new position. The magnetic torque is balanced by the restoring torque produced in the twisted suspension. As the magnetic force acting on the sphere is dependent on the susceptibility of the surrounding gas, the position of the dumb-bell can give the indication of the change in the percentage of oxygen present in the sample gas.

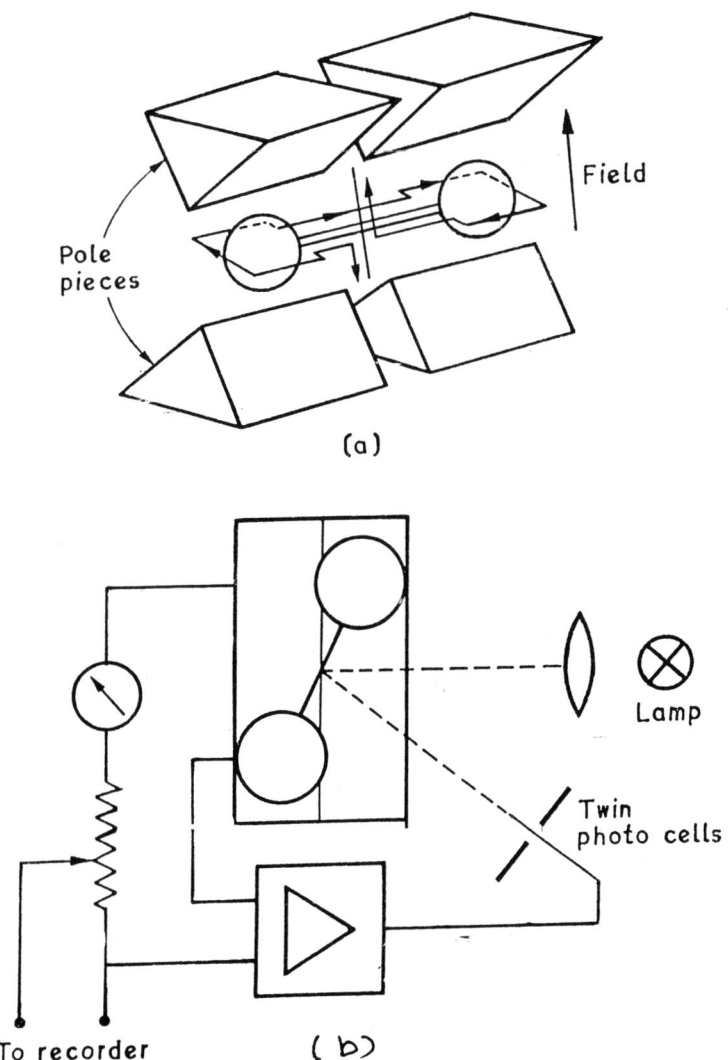

Fig.10.16 : Schematic diagram of a magnetodynamic oxygen analyser (Taylor servomex O_2 analyser is based on this principle).

Fig.10.16 shows, schematically, a magnetodynamic O_2 analyser, designed by Taylor servomex. This is based on Munday cell which differs from Pauling cell in that if uses a platinum ribbon to give a physically strong yet sensitive suspension.electromagnetic feedback is used to supply a neutralizing torque. The position of the dumb-bell is nonitored by a mirror mounted at its centre. This mirror reflects the light from the lamp onto the twin photocells. The difference in the output of the two photocells is amplified and fedback to the coil wound round the dumb-bell. The current required to bring the dumbbell to a zero position is a measure of the oxygen content in the sample gas-flowing through the cell. Such analysers are used for measuring the concentration of oxygen in flue gas.

10.6 CHROMATOGRAPHIC ANALYSERS

In industrial analysis, many situations are encountered where it is required to separate different constituent components from a mixture and analyse them. This job is normally performed with the help of chromatographic analysers. Gas chromatography is employed for isolating compounds in a gaseous phase and is a versatile method of gas analysis. Liquid chromatography has a potential application to nearly all organic analyses.

The chromatographic separation is based on the use of a column (or thin layer) which consists of two phases; viz

   (i) stationary phase and
   (ii) mobile phase.

When the sample is passed through the column, it remains sometime in the stationary phase and rest of the time in the mobile phase. It will move down the column only when it is in the mobile phase. If a mixture of two analytical species (e.g.,two organic compounds) are passed through such a column, the two may spend different times in the stationary phase. The species that spends less time in the stationary phase comes out first where as the other one that spends moretime comes out later and in the process, the two get separated.

The basic components of a process chromatograph are :

   (i)   an inert carrier gas supply with attendant pressure regulators and flow meters,
   (ii) a sample valve for injecting the sample into the carrier flow at set intervals,
   (iii) a separation column or a set of columns with switching facility,
   (iv) an approriate detector and
   (v)   a readout or recorder.

A process chromatograph is shown block diagrammatically in Fig.10.17.

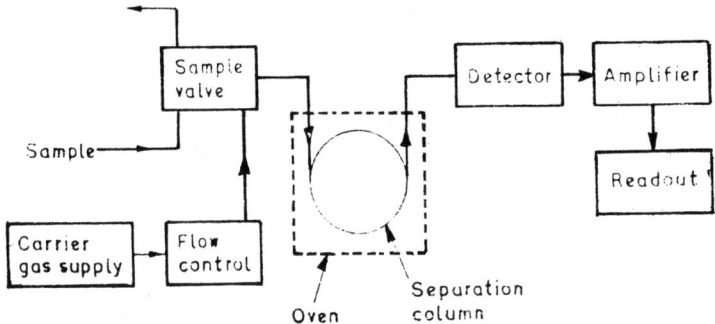

Fig.10.17 : A schematic diagram of a process chromatograph.

The heart of this analyser is the separation column. It consists of a metal tube packed with a support material, e.g. a crushed firebrick. The latter is coated with a liquid substrate. When the gaseous sample passes through the column, different components in the sample are dissolved temporarily for different times in the liquid substrate. Thus the substrate serves as a stationary phase. The support material should be such that it does not react with the sample or the substrate and it should have a high surface area. The substrate should be such that it does not react with the sample or the carrier gas and remain stable at the temperature of operation. Further, it should provide differential partitioning between the constituents of the sample. Carbowax (maximum temperature, $T_{max} = 100°$), castor wax ($T_{max} = 200°$), silicone-greese DC-11 ($T_{max} = 300° C$) etc, are normally used as substrates.

The gaseous sample (or the liquid sample after conversion into the vapour state) is pushed through the packed column with the help of an inert carrier gas, such as helium, nitrogen etc. This carrier gas serves as a mobile phase. In order to achieve reproducibility, the rate of flow of carrier gas must remain constant. The flow rate is normally controlled with the aid of pressure regulators and flow meters.

In a sample valve, the carrier gas flows continuously through one side and the analytical sample (gas or liquid) flows through the sample volume on the other side. At the set time of injection, The sample volume, is moved into the carrier gas side and the gas sample is swept into the packed column. The liquid sample must be vaporized rapidly after injection so that the sample vapour is swept out onto the substrate by the carrier gas. The mechanisms of introducing the sample in the column vary and depend on the type of the sample valve employed in the analyser.

Once the sample is introduced into the column, there is differential partitioning of different components and therefore they emerge out at different time. In order to detect and measure these components, a suitable detector is required at the end of the column. A katharometer or a thermal conductivity detector (TCD) is normally employed for this purpose. These detectors have already been discussed in sec. 10.4. There is a second type of detector called flame ionization detector (FID) which is more sensitive than TCD. In FID, the effluent from the column is fed into the air-hydrogen flame, as shown in Fig.10.18.

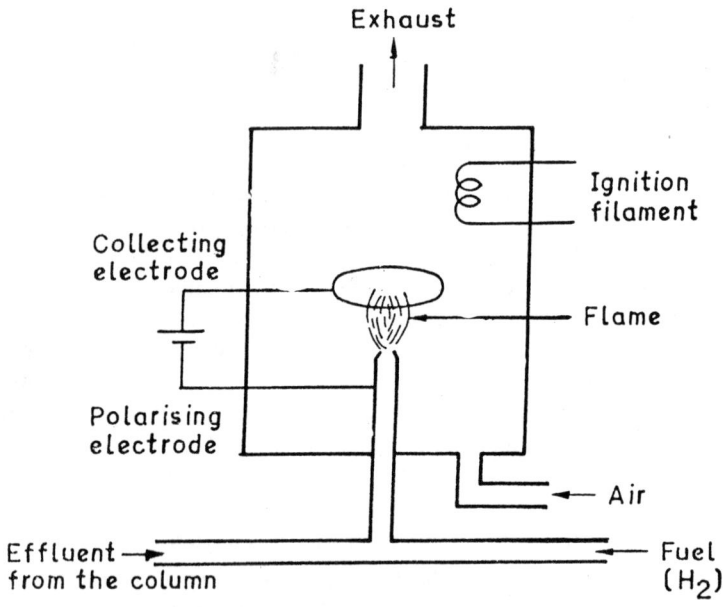

Fig.10.18 : Schematic diagram of a FID

As the sample enters the flame, it is broken up into fragments that are highly conducting. These are detected by the change in current following across the two electrodes. Such detectors are quite sensitive but the sample is destroyed in the process of detection.

The detectors normally measure the change in resistance or conductivity of some part of the detector. Such a change can be easily monitored through Wheatstone's bridge arrangement by using a sample detector in the measuring arm and a reference detector in the reference arm of the bridge. The output of the bridge, current or voltage, is used to drive the pen of the recorder. Thus an electrical signal proportional to the concentration of the sample component passing through the detector is recorded as a function of time. The record is called a chromatogram. The data that can be derived from such a record is disscussed in Tutorial 10.4.

10.7 FLUID DENSITY MONITORS

In some cases of on-line analysis, the measurement of density or specific gravity of a process fluid provides a better means of controlling the product quality.

The density (ρ) of a fluid is defined as the mass per unit volume at a specified temperature. For gases, the pressure should also be specified.

The specific gravity of a liquid is the ratio of the density of the liquid to the density of water at 4°C. The specific gravity of a gas is the ratio of the density of gas to the density of pure dry air at 0°C and 1 atm (760 mm of Hg) pressure.

The density or specific gravity of the fluid can be measured or monitored in a variety of ways. One of the methods employs a float partially submerged in the fluid to be analysed. The device is called a hydrometer. A typical hydrometer is shown, schematically in Fig.10.19. The stem of hydrometer is opaque. If there is any change in the specific gravity of the fluid flowing through the vessel, the position of the stem changes which causes the change in the amount of light falling on the detector. The output of the detector can be calibrated in terms of the specific gravity of the flowing fluid.

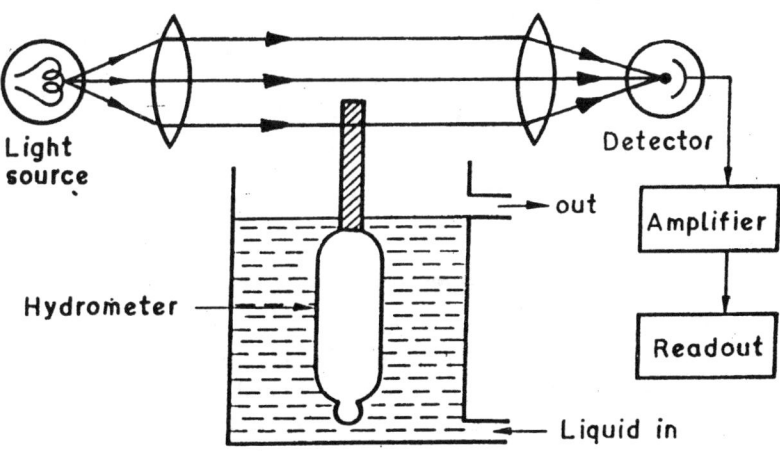

Fig.10.19 : A schematic diagram of a hydrometer for continuous monitoring of specific gravity.

The change in the vertical position of the hydrometer can also be monitored using an inductance bridge as shown in Fig.10.20. Herein, the lower end of the hydrometer is connected to the armature that is free to move through the inductance coil. The variation in specific gravity causes the hydrometer and hence the armature to move up or down. The recording instrument provides an electromagnetic feedback corresponding to the output signal, which in turn provides a balance force. The readout can be calibrated directly in terms of the specific gravity.

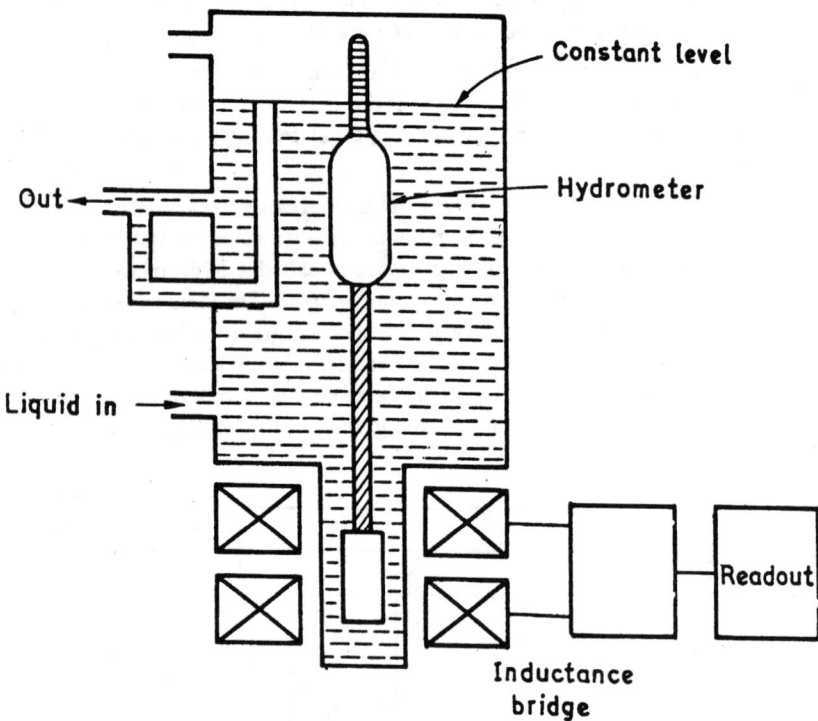

Fig. 10.20: Schematic diagram of an inductance bridge hydrometer.

There is another technique of monitoring the density of liquid that is based on the measurment of pressure difference ΔP between the top and bottom of a column of liquid, as shown in Fig.10.21. If the column height is h and the crosssectional area is A, the pressure difference

$$\Delta P = \frac{\rho Ahg}{A} = \rho hg \qquad (10.6)$$

where ρ is the density of the liquid and g is the acceleration due to gravity. In practice, two pressure transmitters are employed to measure the pressures P_1 and P_2 at two different column heights h_1 and h_2 respectively. If P_a is the atmospheric pressure, P_1 and P_2 are given by:

$$P_1 = \rho gh_1 + P_a$$

and $\quad P_2 = \rho gh_2 + P_a$

Therefore, $\Delta P = P_1 - P_2$

$$= \rho g(h_1 - h_2)$$

$$= \rho hg \qquad\qquad (10.7)$$

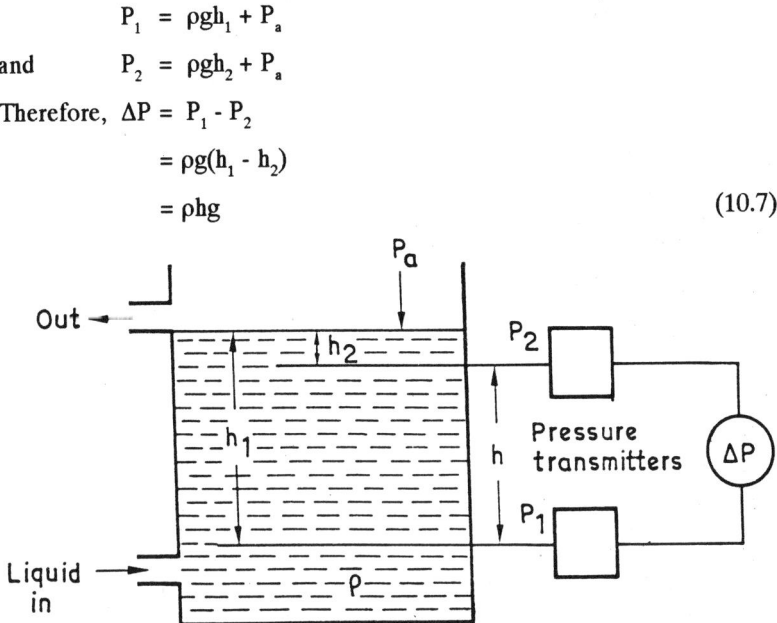

Fig. 10.21 : Measurement of density using a differential pressure head method.

The density of gases can also be measured using a differential pressure head method. A schematic diagram of such a system is shown in Fig.10.22.

The sample gas and the air or reference gas flow through two vertical columns interconnected by a tube. The differential pressure causes the flow through the tube, which in turn produces a differential cooling of the filaments inside the tube. The output of the bridge can be calibrated in terms of of the gas density.

10.8 CONSISTENCY AND VISCOSITY ANALYSERS

The frictional resistance offered by a fluid, when it undergoes a continuous deformation as a shear stress is applied, is called consistency of the fluid. For simple fluids, like water, the consistency is constant if the working pressure and temperature are also constant. If these fluids satisfy certain other requirements also (for example, if they are inelastic) they are referred to as Newtonian fluids. Under these special conditions, the consistency is called viscosity. The viscosity (Ω) for Newtonian fluids is defined as

$$\Omega = \frac{T}{\Gamma} \qquad\qquad (10.8)$$

where T is the shear stress which is the applied force per unit area and Γ is the rate of shear or the velocity gradient produced between the layers of the fluid which is under the shear stress. Ω is offen called dynamic viscosity measured in poise (dyne.sec. cm^{-2}). The kinematic viscosity is given by Ω/ρ and is measured in units of stokes (cm^2 sec^{-1}). Here ρ is the density of the fluid. For non-Newtonian fluids, the viscosity is not constant, but depends on the shear rate Γ.

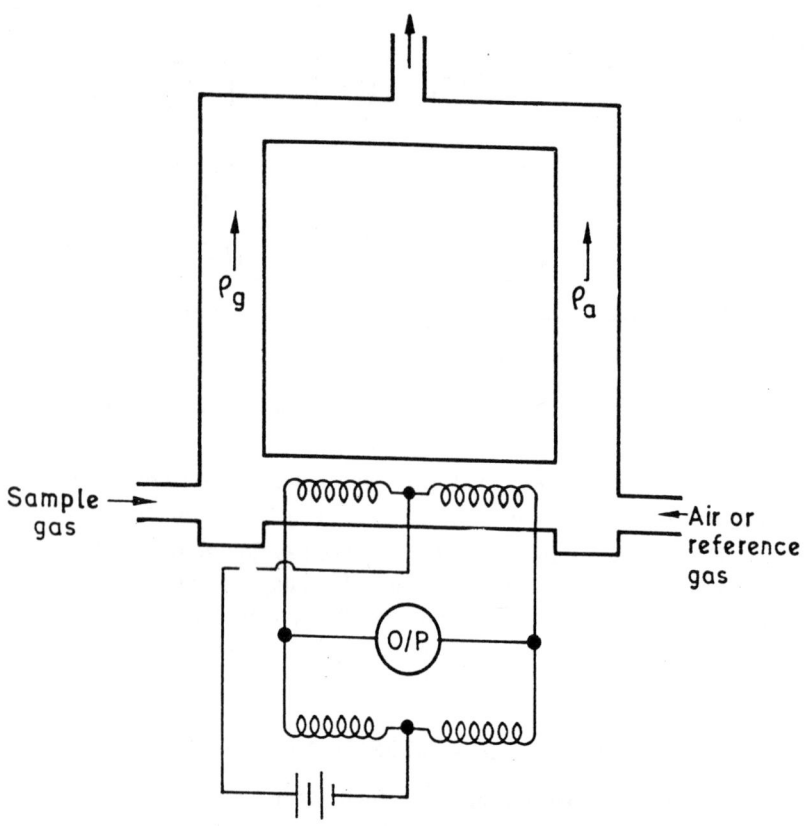

Fig.10.22 : Schematic diagram of a gas Wheatstone bridge

The viscosity or consistency measurement is essential for monitoring various liquid process streams, particularly, those related to the manufacture of paints, paper, heavy oil etc.

A number of on-line viscosity analysers employ viscous flow through a capillary tube for the measurement of viscosity of the fluid. For a laminar flow in a capillary tube the dynamic viscosity of a Newtonian-fluid is given by Hagen - Poiseuille equation,

$$\Omega = \frac{\pi \, \Delta P \, R^4}{8L \, Q} \qquad\qquad (10.9)$$

where ΔP is the pressure difference between the two ends of the capillary tube , R is the radius and L is the length of the capillary, and Q is the volumetric flow rate of the fluid. Thus Ω can be monitored either by measuring the pressure drop ΔP along the tube at constant flow rate or by measuring the flow rate for a constant pressure difference. A schematic diagram of such a viscosity analyser is given in Fig.10.23. Herein, for a constant flow, the pressure drop along the capillary is measured with a differential pressure (ΔP) transmitter. In order to maintain the repeatability, the temperature must be kept constant to within $\pm$ 0.01° C and hence the capillary must be kept in a constant temperature bath. Two or more capillaries may be used in the same bath for analysing different samples simultaneously.

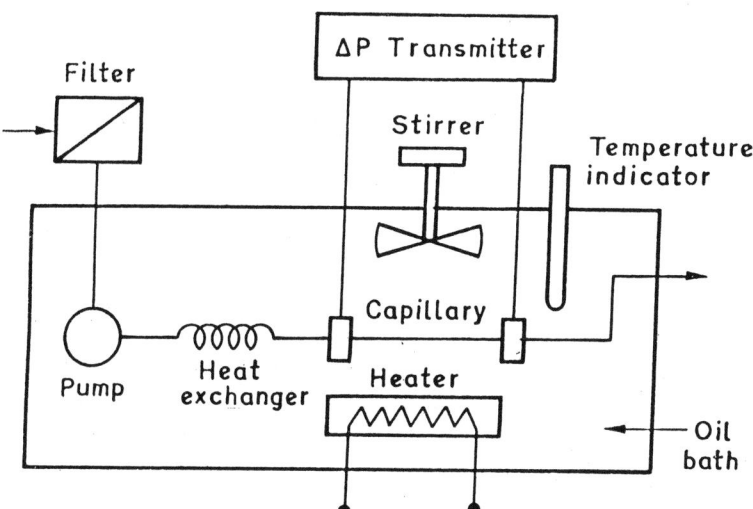

Fig.10.23 A schematic diagram of a capillary viscosity analyser

For measuring the viscosity of non-newtonian fluids, falling body viscometer may be used. It is based on the principle that the velocity V of a body falling under the action of gravity through the fluid is inversely proportional to the viscosity (Ω) of the fluid. A general expression relating V and Ω is as follows:

$$V = \frac{A(\rho_s - \rho_l)}{\Omega} \tag{10.10}$$

where ρ_s and ρ_l are the densities of the solid body and the liquid respectively and A is a constant depending on the position and dimensions of the container and the falling body.

An usual form of the analyser is a solid sphere falling vertically through a viscos fluid in a large diameter cylinder. An automatic operation is achieved by monitoring the position of the falling sphere through electromagnetic coils around the cylinder. A typical falling sphere viscometer is shown in Fig.10.24. In this case, the solenoid valve (SOV) is opened when the steel ball reaches the bottom of the cylinder. The opening of SOV raises the ball upwards and also replaces the old sample by a new one. SOV is closed when the ball reaches the top. The viscosity signal is obtained by measuring the time of fall of the ball as it passes from coil 2 to coil 3.

There is another variety of viscometers that is based on the measurement of viscous drag force experienced by the measuring element put inside the viscous fluid. These are used in measuring the consistency of the paper-pulp stock and that of a syrup-crystal mixture in a sugar factory.

10.9 APPLICATIONS OF ON-LINE ANALYSERS

On-line analysers are an integral part of nearly all the continuous proces plants. Their role is manifold. The first role is that of a controller of the process so that it can be run within safe limits and at optimum efficiency. The job of an analyser here is to maintain the product quality. Thus it may be required to check the liquid or gas composition, setting or resetting of some parameters; e.g. flow rate, temperature etc; measuring some physical property and so on. Depending on the requirement, therefore, a gas chromatograph, TC, NDIR or paramagnetic oxygen analyser may be used.

Analysers are also required for monitoring the concentration of some gases in the specific locations. The concentration of the gas itself may not be of direct interest but the monituring may give the indication or the warning of an event. For example, the combustible gases and vapours may propagate a flame if their concentration in air is not within the explosive limits. If the atmosphere containing such vapours is ignited, an explosion may result. Thus, in plants like refineries, petrochemical and chemical works or gasworks, gas detecting analysers are very common.

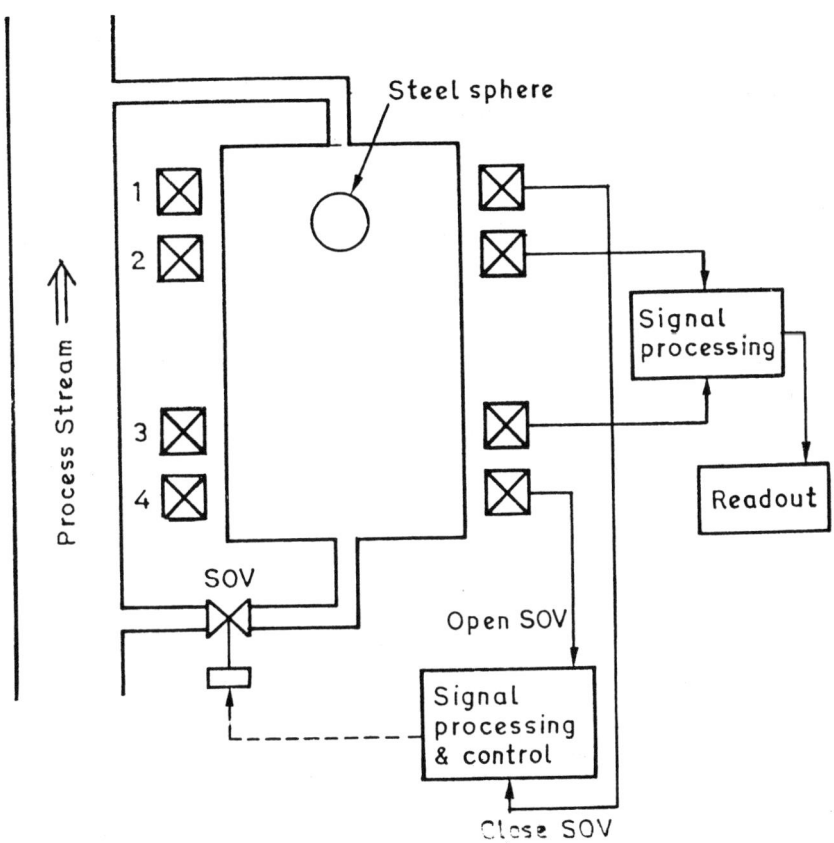

Fig.10.24 : A schematic diagram of a falling sphere viscometer (1,2,3 & 4 are electromagnetic coils),

Further, there is always danger of plant personnel being exposed to the gaseous releases in minute quantitites. Concentrations of the order of 100-1000 ppm of toxic gases may produce harmful effects. Thus guidelines have been framed with regard to the maximum allowable concentration of nearly all the gases and vapours to which the workers may be exposed. The analysers are required to monitor the concentration of these gaseous releases in the plant area. Analysers are also used for controlling the air pollution caused by the gaseous releases in the open atmosphere.

TUTORIAL-10

10.1 In order to monitor the environment e.g., detection of combustible gases, it is desirable to use a multipoint sample system but from the economic point of view, it is not possible to use separate analysers for all sampling points. Suggest (a), the design of sample system for environmental monitoring at a single point, and (b) how this design may be modified to incorporate a number of points using the same single analyer.

10.2 Most of the flue gas analysers are required to monitor oxygen, CO or CO_2 and sometimes SO_2 and NO_x as well. The flue gases are generally hot, wet and dirty also. Suggest a sampling system for such analysers.

Hint: For sampling flue gas, water washed probe with a water ejector would be most suitable. This should be followed by a coarse bypass filter. Other modules remain as usual. However, this system will not work, if a water-soluble gas is to be analysed. In this case, a filter-probe followed by a cooler would be needed.

10.3 Suggest a method for measuring the concentration of a particular gas component (say,A) in a mixture of two gases (say, A+B) employing a TC analyser.

10.4 A chromatogram of a typical sample (gas) was recorded using a liquid substrate in a chromatography column. Relevent data is as follows.

Temperature of the column	= 50°C
Length of the column	= 100 cm
Amount of liquid substrate in the column	= 3.5 gm.
Density of the substrate	= 0.95 gm/cm³
Chart speed of the recorder	= 3 cm/min
Flow rate	= 60 ml/min
Uncorrected retention time	= 70 cm (chart)
Retention time for air	= 10 cm (chart)
Carrier gas pressure at the inlet of the column	= 120 cm.
Carrier gas pressure at the outlet of the column	= 80 cm
The basewidth of the peak	= 6 cm.

Calculate

(a) the adjusted retention time,
(b) the adjusted retention volume,
(c) the net retention volume,
(d) the specific retention volume at the temperature of the column,
(e) the efficiency of the column in terms of the theoretical plates, and
(f) HETP of the column.

Answers: (a) 20 min, (b) 1200 ml, (c) 947.36 ml, (d) 228.37 ml, (e) N = 1600 (f) H ETP = 0.75 mm.

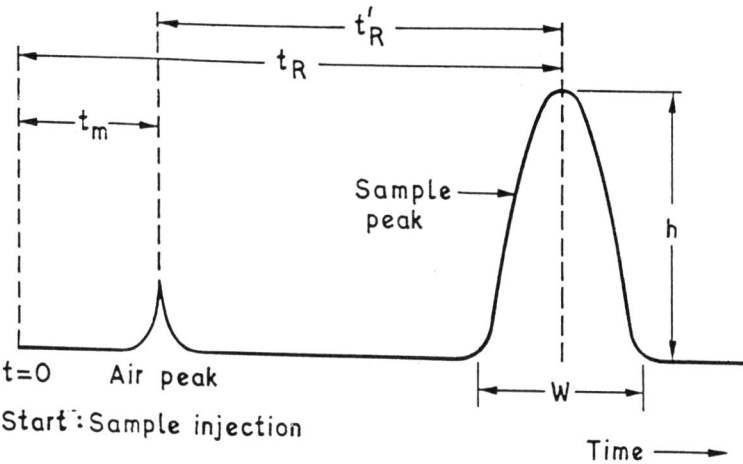

Fig.10.25 : A typical chromatogram

Hint :

(a) The retention time, t_R, is the total time spent by the sample in the chromatography column; i.e., the time spent in the mobile (gas) phase and the stationary (liquid) phase. The time spent in the mobile phase is known by injecting air prior to the sample. Assuming that air does not interact with the substrate, and therefore, it flows down the column in a minimum of time. This period is called dead time, t_M. It is a measure of the time taken by the sample to flow through the column without interacting with the substrate.

   The adjusted retention time, $t'_R = t_R - t_M$ is thus a measure of the extra time the sample spends in the column. In fact, this is the time spent by the sample in the stationary phase. The relevent paramters are shown in fig. 10.25.

(b) The retention volume (V_R) of the sample is defined as the volume of the carrier gas that flows through the column during the retention time (t_R). Thus, if the flow rate of the carrier gas is F_c, it may be given by

$$V_R = t_R \cdot F_c.$$

   The adjusted retention volume, $V'_R = t'_R \cdot F_c.$

(c) As the mobile phase is compressible, a pressure gradient correction must be applied to V'_R to get the net retention volume, V_N. Thus

$$V_N = jV_R$$

where $\quad j = \dfrac{3[(P_i/P_o)^2-1]}{2[(P_i/P_o)^3-1]}$

j is called a compressibility factor; P_i and P_o are the gas pressures at the inlet and outlet of the column respectively.

(d) The specific retention volume,

$$V_g = \dfrac{273}{T_c} \times \dfrac{V_N}{w_L}$$

where T_c is the absolute temperature of the column and w_L is the weight of the liquid substrate in the column.

(e) The efficiency of the chromatography column is measured in terms of the number of theoretical plates, which is given by

$$N = 16 \left(\dfrac{t'_R}{W}\right)^2$$

where W is the base width of the peak.

(f) The height equivalent of the theoretical plate,

$$HETP = \dfrac{L}{N}$$

where L is the length of the column.

10.5. **pH - Meter:** A large number of analysers cum controllers have been developed which are based on the measurement of pH. This method employs special electrodes which develop an emf proportional to the hydrogen ion concentration in the solution into which they are immersed. The pH of the solution is defined as:

$$pH = - \log_{10} [\text{ concentration of } H^+, \text{moles/l }]$$

$$= \log_{10} \left[\dfrac{1}{\text{Concentration of } H^+, \text{moles/l}} \right]$$

The chemical activity of all the aqueous solutions of acids and alkalies is due to the relative concentration of H^+ and OH^- ions. In pure water the equilibrium product of the concentration of H^+ and OH^- ions is constant and is equal to 10^{-14} at 25°C. Thus the concentration of H^+ in pure water is 10^{-7} and hence the pH = $\log (1/10^{-7})$ = 7.0. If the pH of the sample solution is less than 7, it is acidic and if the pH is greater than 7, it is alkaline. The advantage of this method is that all values of acidity and alkalinity between these, of solutions molar with respect to hydrogen ions, can be expressed by a series of positive numbers between 0 and 14.

Suggest the design of an electrochemical analyser based on monitoring the pH of the solution in a continuous process stream.

10.6 **Cloudpoint and pour point analysers:** If a hydrocarbon oil is cooled gradually a temperature is reached at which the wax starts to precipitate from the oil. This temperature is called the cloud point. If this oil is cooled further, it solidifies and pours under its own weight. This temperature is called the pour point.

Suggest the design of analysesrs that is based on absorption of light for detecting the cloud point and pour point of a hydrocarbon oil.

10.7 **Flash Point Analyser :** The flash point is a measure of the suitability of the fuel. It is the temperature at which the liquid (fuel) gives off sufficient vapour to form an ignitable mixture with air. The ignition source may be either a flame or spark.

In order to determine the flash point of the fuel, its sample is pretreated in the following way. The sample is cooled sufficiently below the flash point and then passed through a coalescer to remove the water in suspension. The treated sample is then heated again upto a temperature of flash point, for a period of about 2 minutes. At this point, the liquid fuel ignites. This ignition can be detected by increase in, either temperature or pressure of the liquid vapour.

Suggest a method of recording the flash point temperature of the fuel.

10.8 Suggest a design of an explosimeter for detecting flammable or combustible gases in the plant environment.

10.9 The reaction of the liquid sample with a suitable reagent and further processing of the reaction product (e.g., thermal treatment) may give rise to the colour. This coloured product may then be used to identify the analytical species (i.e. the liquid sample) through colorimetry or some other optical method.

Suggest an instrument design, based on this principle, for automatic analysis of liquid process streams.

SELECTED BIBLIOGRAPHY

1. Bauman, R.P., "Absorption spectroscopy", John Wiley, New York (1962).
2. Chappell, A. (Ed). "Optoelectronics: Theory and Practice" Mc.Graw Hill, New York (1978).
3. Considine, D.M. (Ed.), "Process Instruments and Controls Hand Book", Mc Graw Hill, New York (1974).
4. De Vany, Arthur, S. "Master Optical Techniques", John Wiley, New York (1981).
5. Driscoll, W.G. and Vaughan, W. (Eds), "Hand Book of optics", Mc.Graw Hill, New York (1978).
6. Edisbury, J.R., "Practical Hints on Absorption Spectrometry", Adam Hilger Ltd., London (1966).
7. Ewing, G.W. "Instrumental Methods of Chemical Analysis", (4th Ed.), Mc.Graw Hill Kogakusha Ltd., Tokyo (1975).
8. Grove. E.L. (Ed) "Analytical Emission Spectroscopy" Vol. I, Part I & II, Marcel Dekker Inc., New York, (1971).
9. Honig, J.M., and Rao, CNR (Eds), "Preparation and Characterization of Materials", Academic Press, New York (1981).
10. Huskins, D.J., "Quality Measuring Instruments in On-line Process Analysis", Ellis Horwood Ltd, Chichester (1982).
11. Jones, E.B., "Instrument Technology" Vol.2, Newnes-Butterworths, London (1976).
12. Khandpur, R.S. "Handbook of Analytical Instruments", Tata McGraw Hill, New Delhi (1989).
13. Kingslake, R.(Ed).,"Applied Optics and Optical Engineering",Academic Press, New York, Vol. I (1965) and Vol.IV. (1967).
14. Koller, L.R., "Ultraviolet Radiation" (2nd Ed), John Wiley, New York (1965).
15. Lawes, G. "Scanning Electron Microscopy and X-ray Microanalysis"(ACOL) John Wiley, Chichester (1987).
16. Levi, L. "Applied Optics", John Wiley, New York (1960)
17. Malacara, D.(Ed), "Optical Shop Testing", John Wiley, New York (1978).
18. Mann, C.K., Vickers, T.J. and Gulick, W.M. "Instrumental Analysis" Harper

& Row, New York , (1974).

19. Mavrodineau, R. (Ed)"Analytical Flame Spectroscopy" MacMillan, London (1970).

20. Mika, J. and Torok, T. "Analytical Emission Spectroscopy, Fundamentals", English Edition, Butterworths, London (1974).

21. Morrison, G.H. (Ed.) "Trace Analysis, Physical Methods", Wiley Interscience, New York (1965).

22. Noltingk, B.E. (Ed), "Jones' Instrument Technology" (Fourth Edition) Vol 1,2 & 3, Butterworths, London (1985).

23. Potts, W.J., "Chemical Infrared Spectroscopy" Vol. I, John Wiley, New York (1963).

24. Richardson, J.H. and Peterson, R.V. (Eds), "Systematic Materials Analysis", Vols. I,II,III & IV, Academic Press, New York (1974).

25. Robinson, J.W. "Undergraduate Instrumental Analysis" (3rd Ed.) Marcel Dekker Inc. New York (1982).

26. Sawyer, R.A. "Experimental Spectroscopy"(2nd Ed.), Prentice Hall, New York (1951).

27. Skoog, D.A. "Principles of Instrumental Analysis" (3rd Ed.) Holt-Saunders, Philadelphia (1985).

28. Strobel, H.A., "Chemical Instrumentation: A Systematic Approach",(2nd Ed.), Addison Wesley, Reading, Massachusetts, (1973).

29. Thornton, P.R., "The Physics of Electroluminescent Devices", E & F.N. Spon Ltd., London (1967).

30. Verdin, A., "Gas Analysis Instrumentation", MacMillan, London (1973).

31. Whiston, C."X-Ray-Methods" (ACOL), John Wiley, Chichester, (1987).

32. Willard, H.H., Merritt, L.L.; Dean, J.A. and Settle, F.A., "Instrumental Methods of Analysis" (6th Edition) CBS Publishers, New Delhi (1986).

33. Williams, D.A.R. "Nuclear Magnetic Resonance Spectroscopy" (ACOL), John Wiley, Chichester (1986).

INDEX

208 Analysis Instrumentation an Introduction